NATO SERIES

ATLANTIC COUNCIL OF THE UNITED STATES

NATO AND THE NEW TECHNOLOGIES

David Hobbs

UNIVERSITY PRESS OF AMERICA

Lanham • New York • London

University Press of America,® Inc.

4720 Boston Way
Lanham, MD 20706

3 Henrietta Street
London WC2E 8LU England

Printed in the United States of America

British Cataloging in Publication Information Available

Co–published by arrangement with
The Atlantic Council of the United States

Library of Congress Cataloging-in-Publication Data

Hobbs, David, 1956–
NATO and the new technologies / David Hobbs.
p. cm.
(NATO series)
"Co–published by arrangement with the Atlantic Council of the United States"– –T.p. verso.
1. North Atlantic Treaty Organization– –Armed Forces– –Weapons systems. 2. North Atlantic Treaty Organization– –Armed Forces– –Procurement. I. Atlantic Council of the United States. II. Title. III. Series.
UA646.3.H58 1989 88–39010 CIP
355.8'2'091821– –dc19
ISBN 0–8191–7323–1 (alk. paper).
ISBN 0–8191–7324–X (pbk. : alk. paper)

All University Press of America books are produced on acid-free paper.
The paper used in this publication meets the minimum requirements of American National Standard for Information Sciences—Permanence of Paper for Printed Library Materials, ANSI Z39.48–1984.

For my Parents

Acknowledgements

I owe debts of gratitude to many people who have helped me to produce this book. Foremost among them are Greg Flynn and Stan Woods, both then at the Atlantic Institute in Paris, who—in my view—showed excellent judgment in identifying the need for a short yet comprehensive survey of new military technology ("Everyone is talking about it but no one understands it") and which placed military technology into a procurement context. They showed rather more questionable judgment in proposing that I write it, but I am extremely grateful that they did so. Greg Flynn deserves an extra vote of thanks for his very useful comments and suggestions on successive drafts. I would also like to thank the Atlantic Council of the United States and particularly Job Dittberner and Joseph Harned for their assistance and support.

Intellectual debts to Mike Moodie, David Greenwood, and Schuyler Foerster must also be acknowledged. Discussions with them on many issues covered in this book were invaluable. Thanks also to John Borawski who read the (almost) final draft and helped me to straighten many formerly convoluted points.

Marie-Claire Schockaert did a magnificent and much appreciated job in typing the numerous drafts. That she can read my writing is impressive, that she can read it quickly is amazing. And many thanks to Carolyn Butler and Vanessa Noble who stepped in during emergencies.

Last, and certainly not least, thanks to my wife, Bernadette, whose knowledge of grammar, proof reading abilities and—above all—moral support made this possible.

CONTENTS

LIST OF TABLES

FOREWORD

This book is an important contribution to understanding the impact of technological change and innovation on Western defense, Alliance cohesiveness and deterrence options. Author David Hobbs takes us beyond speculation about the impact of technological change and manages with great thoroughness and clarity to survey the technological "menu" now facing the Alliance, to analyze the deficiencies of new high-tech options, and to suggest some methods of correcting these deficiencies.

This volume is a serious contribution to our understanding of future Alliance options for two additional reasons more political in nature. The first concerns maintaining conventional forces adequate, in combination with our nuclear forces, to preserve a high level of deterrence. One result of the INF Treaty has been increased attention to the imbalance of conventional forces between NATO and the Warsaw Pact. High-tech weapons systems may indeed offer advantageous enhancement to the conventional dimension of our deterrent power, and compared to alternatives may be relatively less costly. However, the cost and the way the members of the Alliance procure new weapons brings us to the second consideration.

In a period of constrained resources for all the NATO countries, the Alliance can no longer afford the luxury of uncoordinated national policies if it is to remain strong in deterrence and defence. Differing national approaches to the design and suitability of weapons systems have not only strained relations among allies in the past, but have at times distracted member nations from the common task of a collective defence. David Hobbs' study is particularly timely because he examines the possibility of adoption of new technologies in the context of an issue that has consistently plagued the NATO Alliance: harmonization and coordination of national means of procurement and production. Such coordination of national policies is critical if the Alliance is to obtain the most value for money spent on weapons, equipment and services.

While training, equipping and maintaining combat-ready forces is a national responsibility, the fact remains that in order "to exploit available technologies at an affordable cost NATO clearly must organize its defense procurement more efficiently.'' We must learn and embrace a common understanding of the relationships between quantity and quality, and we must be able to strike an appropriate balance between the two—a balance that places a high priority upon both reliability and ease of maintenance. The Alliance, as the author concludes, must ''pay greater attention to first defining and then meeting Alliance force requirements instead of essentially merging national ones.''

While the views in this book are, of course, solely those of the author and should not be ascribed to the Directors or staff of the Atlantic Council of the United States, the Atlantic Council is pleased to have been able to support this study, as it reaffirms the Council's long-standing commitment to a stable and viable NATO Alliance.

On behalf of the Council, it is my pleasure to thank those who have supported this and related projects and activities—the Andrew W. Mellon Foundation, the Pew Charitable Trusts, and the John D. and Catherine T. MacArthur Foundation. Their generous support has been essential in continuing the Council's educational activities as well as the policy studies aimed at maintaining wise U.S. leadership in an increasingly interdependent world.

George M. Seignious II
President
Atlantic Council of the United States

EXECUTIVE SUMMARY*

NATO has traditionally relied on its technological superiority to offset the Warsaw Pact's numerical advantage. This equation, however, is becoming increasingly difficult to sustain. Despite NATO's considerably greater outlays for defense, the numerical balance continues to favor the Warsaw Pact and the technological "gap" between East and West is narrowing. To rectify this situation, NATO has two courses of action. One is to embrace a spectrum of new technologies which offer the prospect of dramatic improvements in capability. Technological progress in many fields is opening up promising new mission possibilities as well as enhancing the performance of traditional weapons systems. This situation is therefore giving rise to a debate about how best to exploit these new technologies and to what degree NATO should rely upon a variety of innovative systems.

NATO's other course of action is to extract better value for money from its defense allocations. There is ample scope for eliminating wasteful duplication of effort and other procurement inefficiencies but the problems tend to be easier to identify than to resolve. Solutions must be found, however, because there is no prospect in the immediate future of any substantial increases in funding for defense. On the contrary, current trends are towards level funding or even a decline in real terms.

These two possibilities—better value for money and the vigorous adoption of new technologies—need not in principle be mutually exclusive but historically they have been. As weapons systems have become increasingly sophisticated, several key trends have emerged. The research and development investments necessary to produce weapons systems have risen to represent a substantial proportion—up to 30 percent—of total program costs; acquisition and operational costs have

* Much of this book is devoted to descriptions of "leading edge" technologies and explanations of how they will or could be employed. Such descriptions and explanations do not lend themselves to useful condensation so this summary concentrates on the book's analysis of the weapons procurement process within NATO and how that process should be modified.

risen dramatically and often unpredictably; programs have been stretched to keep outlays within annual budgetary provisions so that production lines have operated at less than optimal rates causing further increases in unit costs. The net effect of all this is to reduce weapons inventories.

Comparing NATO and Warsaw Pact Procurement

Clearly NATO must reform its approach to weapons procurement if it is to exploit new technology as effectively as possible in a period of budgetary stringency. The need for reform is evident: the Warsaw Pact spends less than NATO on defense but outproduces NATO in virtually every weapons category. So how does Warsaw Pact procurement differ from NATO's and enable weapons to be produced with such efficiency?

The essential features of Warsaw Pact procurement structure are:

1. The Soviet Union is overwhelmingly the principal Warsaw Pact weapons producer which automatically leads to the benefits of very long production runs, economies of scale, and a high degree of standardization.
2. Central planning accords defense procurement an extremely high priority so that new technology is developed (or acquired) and applied as necessary.
3. Weapons development is evolutionary rather than revolutionary so that even "new" weapons incorporate tried and tested—and possibly incrementally improved—technology yielding reduced research and development costs and benefits in areas such as ease of maintenance, maintenance training, reliability, interoperability, and logistics.

For NATO, the essential features are:

1. Several significant national producers with relatively short production runs, paying little attention to standardization in programs which are often duplicative.
2. Substantial interplay between national defense procurement policies and economic policies such that procurement decisions are influenced by considerations such as employment, balance of payments, and maintenance of national technological and industrial bases.

3. A largely revolutionary procurement philosophy which emphasizes weapons performance parameters such as speed, range, agility etc. rather than ease of maintenance, reliability, and interoperability.

It would be naive to suggest that NATO should or could model its procurement on the Warsaw Pact's but the Warsaw Pact structure does at least demonstrate the principle that procurement can be more efficient and some useful lessons are suggested by comparing each procurement strategy.

Cooperation and Competition in Procurement

Compounding the problems of structural disarmament—the process whereby increasing sophistication results in reduced weapons inventories—is the large scale duplication of procurement effort within NATO. Many nations produce essentially similar systems so that resources for research, development, and production are squandered duplicatively and each nation's investment must be amortized over a relatively short production run. This situation has never been ideal but the problem has worsened as research and development costs have become more significant while unit costs have also risen substantially. As a result, there is increasing recognition of the need to coordinate defense procurement in cooperative programs so that costs can be shared and amortized over longer production runs. In other words, the benefits of economies of scale must be realized.

Defense cooperation comes in many forms, however. Divisions of labor on a particular program represent the most popular and probably least cost-effective arrangement. Other options which ought—in principle at least—to offer greater economies are project packaging and cross purchasing. Project packaging essentially involves formulating a division of labor over several programs to provide more flexibility in assigning program segments. It enables, for instance, a participant to opt for a small share of one program in return for a larger share of another. Cross purchasing is basically a barter arrangement whereby nations exchange items of defense equipment. This avoids duplicative production and allows both parties to realize economies of scale.

Such arrangements can be difficult to negotiate because procurement schedules for a number of systems must be coordinated and concerns about losing national competence can arise. Even so, some successes have been achieved and greater attention should be paid to identifying opportunities for these modes of cooperation.

Greater attention should also be given to increasing competition in defense production. Efforts are being made to open up defense contracts to national and even international competition but still only about 25 percent of NATO procurement by value is open to competitive tendering and much of that on a strictly national basis. Even so, competition has yielded considerable price reductions in many systems and the case for opening up more contracts is clear.

Operations and Maintenance Costs

Reducing acquisition costs is only one way of relieving pressure on defense budgets. Equally important is the need to reduce operations and maintenance costs. Cooperative procurement goes some way towards this by increasing standardization and interoperability but procurement philosophy must also specifically address reliability and ease of maintenance in order both to reduce operations and maintenance costs and to increase operational effectiveness.

In the past, growing weapons complexity has led to decreasing reliability and a need for more maintenance personnel, spare parts, and test equipment. Furthermore, these burdens have often been much larger than anticipated so defense budgets have had to accommodate inescapable but unforeseen and recurring additional costs.

To avoid these problems in the future NATO must embrace a new philosophy regarding weapons technology. This involves a rearrangement of the yardsticks of performance. Reliability and maintainability must rank equally with more traditional parameters such as range, speed, agility, and "smartness". At the same time, the tendency to pursue "system-for-system" replacement practices must give way to a "mission-for-mission" approach. New technology is extremely versatile and offers many opportunities for conducting new missions and for conducting old missions in new ways. Options should not therefore be constrained by—for example—service pressure to substitute new systems for old when other, more cost-effective opportunities present themselves.

Nor should systems be chosen simply on the basis of technological virtuosity. New technology should be applied as appropriate rather than as available. In other words, the relationship between technology and doctrine should be symbiotic: technology should not compel the adoption of inappropriate doctrine nor should out-dated doctrine constrain technology. To facilitate this symbiosis and to ensure operational effectiveness, increased emphasis must be placed on rigorous and realistic testing at all suitable stages in the procurement process. Simulation is a valuable tool but over reliance on simulation at the expense of thorough testing is a false economy.

Transatlantic Differences and the European Pillar

Technology issues have a substantial bearing on the United States' relationship with its Allies. The United States enjoys a larger domestic market and a substantial foreign market which support a military research and development budget over three times as large as collective European research and development budgets. Such a difference in research and development funding is disadvantage enough without the added problems of duplication in research, development and production combined with relatively small national markets. Ideally, all Alliance research and development efforts should be coordinated but the need is greatest in Europe. European research and development efforts should be harmonized more effectively not just for the obvious economic reasons but also for Europe to avoid technological dependence on the United States, to be competitive in the American marketplace, and to be treated as an equal partner in transatlantic cooperative high-technology projects. For the same reasons, while enhanced cooperation and competition is needed throughout the Alliance, they must be promoted within Europe with particular urgency. Movement towards greater European cooperation—by such methods as mutual opening of markets, joint procurement projects and perhaps the creation of some form of common research and development funding—should not be seen as "anti-American" but rather as a means of enhancing the Alliance's defense production output and, therefore, the Alliance's security. This should also help to reduce criticisms of European defense efforts. There is little prospect of Europe increasing the resources devoted to defense—its defense *inputs*—but it could increase its defense *outputs* and these are, after all, a more important measure of security.

Technology transfer issues are another source of transatlantic tension which is now being addressed more constructively but which still merits attention. The NATO nations and other countries in the Western camp operate technology export controls to hamper the Soviet Union's substantial efforts to acquire militarily relevant technology. Unfortunately, these controls have—albeit inadvertantly—constrained technology transfer within the Alliance. The source of this problem is the "two-tier" export control framework consisting of controls imposed uniquely by the United States and those imposed multilaterally by all Alliance nations (except Iceland) plus Japan through the Coordinating Committee on Multilateral Export Controls (COCOM). Great progress is being made in removing the procedures which hamper technology transfer from the United States to its COCOM allies and this momentum should be maintained in order to drain the reservoir of ill will which still persists on this matter within the Alliance. Efforts to harmonize export controls should therefore continue and greater attention should be paid to involving industry in the export control process and to taking account of foreign availability of controlled technology.

Political Constraints on New Technology

In operational terms, new technology obviously has profound consequences. Enhanced communications and navigation capabilities will, for instance, greatly improve the effectiveness of operations involving coordinated action by air, sea, and land forces thus augmenting NATO's defense of the Northern and Southern Regions in particular. NATO does not, however, have a completely free hand in embracing new technology. Political considerations cannot be neglected even though in strictly military terms these can appear misguided and even capricious. The notion of the counter-offensive, for example, can arouse political opposition sufficient to delay or even preclude the acquisition of certain capabilities which can be perceived as provocative and deployment of nuclear-capable missiles of any variety cannot be considered in Europe without careful evaluation of the political consequences. Similarly, movement towards a more mobile conventional defense strategy facilitated by several new technologies could meet political resistance because it implies the acquisition of some counter-offensive capability as well as the prospect of yielding some terrain without a struggle.

Yet, paradoxically, the notion of pre-positioning fixed defenses such as advanced mines in order to bolster forward defense raises opposition on the basis that it would symbolize political acceptance of the division of Europe.

These obstacles can in some cases be circumvented by opting for somewhat less cost-effective and militarily less efficient alternatives. Examples include air-launched and air-delivered systems instead of ground-based missiles since aircraft do not seem to arouse as much public hostility, and rapidly deployable mines instead of pre-positioned ones. Ultimately, however, difficult choices will have to be faced. Those nations whose forces share the task of defending the East-West border, and whose economies must shoulder that burden, cannot be expected indefinitely to make the operational compromises and economic sacrifices necessary to mollify inconsistent public and political opinion. Consequently, there is a clear need to present operational and equipment choices more constructively and clearly than has been necessary in the past. That includes presentation of the alternatives such as increases in defense expenditure, reductions in security, and a diminution of Allied forces.

Conclusion

To exploit available technologies at an affordable cost NATO clearly must organize its defense procurement more efficiently. That must involve defining and meeting Alliance force requirements instead of merely merging national ones. Furthermore, those requirements should be formulated in accordance with a mission-oriented philosophy which accords a high priority to reliability and ease of maintenance. NATO has recognized the need for procurement reform and is making cautions steps in the right direction. Having established the direction to go, NATO must now proceed more rapidly and boldly towards more rationalized procurement. Inefficient research, development and production are luxuries the Alliance can no longer afford.

INTRODUCTION

Numerical assessments of Warsaw Pact and NATO forces show a clear Warsaw Pact superiority but not, at present, an unambiguous military advantage. Defending forces can prevail against a numerically superior aggressor and NATO's overall technological superiority helps to off-set its quantitative inferiority. Furthermore, NATO's capacity and stated intention to resort to nuclear forces if or when its conventional defenses begin to crumble are intended to pose sufficient risk to render a Warsaw Pact onslaught an unlikely eventuality.

The situation, however, is changing in several dimensions. The Warsaw Pact's numerical advantage is increasing and the quality of its military equipment is improving, such that NATO's technological superiority is being eroded. In addition, movement towards reductions in nuclear weapons inevitably heightens the significance of disparities in conventional forces. NATO has in recent years unilaterally reduced its arsenal of short-range nuclear forces, superpower nuclear disarmament negotiations produced in December 1987 a treaty eliminating intermediate-range nuclear forces, and further superpower negotiations are likely to carry nuclear disarmament still further.

Consequently, one of the key challenges facing NATO is how to maintain conventional forces in sufficient quantity and of sufficient quality to deter Warsaw Pact conventional aggression and, if deterrence fails, to rebuff an attack without having to resort rapidly to nuclear weapons for want of the means to mount a robust conventional defense. This challenge is not, of course, a purely military one; a substantially adverse *military* balance could leave Western Europe and NATO potentially vulnerable to Soviet attempts at *political* intimidation.

To achieve adequate quantity and quality, NATO will have to exploit new military technology. Military technology is constantly advancing, offering improvements in military systems and opening up new mission opportunities. Unfortunately, entrenched practices and political obstacles bedevil NATO weapons procurement and call into question its ability to exploit its technological capacity to the full.

That NATO has the will and ability to succeed in overcoming these difficulties is by no means assured, despite the obvious urgency of doing so. As one analyst has commented: "Emerging technology has an enormous potential to contribute to the improvement of NATO's conventional defense. Unfortunately, the Alliance record offers marginal hope of achieving a fraction of that potential".[1] The purpose of this study is to describe this technological "menu" now facing the Alliance, analyze deficiencies in the Alliance's capacity to reap the benefits of new technology, and suggest some methods of correcting these deficiencies.

The first chapter sets the scene by describing NATO's current procurement practices, highlighting the deficiencies, their causes, and their consequences. Following that are four chapters describing the new technologies now in prospect. These are categorized under the headings "Electronics Technology", "Battlefield Technology", "Aerospace Technology" and "Naval Technology". Each chapter describes the relevant new technologies and provides examples of systems in development based on those technologies although inevitably these headings overlap to some extent and the locations of some items are necessarily arbitrary. The final chapter assesses the steps being taken to improve weapons procurement processes and to avoid—or at least to minimize—the operational problems which have consistently plagued high technology weapons. This chapter analyzes issues such as the relationship between doctrine and technology, and quality versus quantity and proposes some guidelines on how NATO can ensure that it takes full advantage of the new technologies now in prospect. It also discusses the political constraints on employing certain new technologies and appraises the effects of new technology on operations in the Northern and Southern Regions, areas often neglected in considerations of the impact of technological progress.

1

NATO's PROCUREMENT PROBLEMS

The shortcomings of NATO weapons procurement and the urgency with which these must be addressed are clearly illustrated by comparing NATO and Warsaw Pact weapons procurement.

Warsaw Pact Procurement

The Warsaw Pact enjoys numerical superiority over NATO in virtually every weapons category and continues to out-produce NATO. (See Tables 1 and 2). Yet, the Warsaw Pact achieves this while spending less on defense. For instance, according to figures cited by the International Institute for Strategic Studies, the Warsaw Pact spent about $259 billion on defense in 1983, while NATO spent about $343 billion,[2] a remarkable situation even allowing for substantial differences in manpower costs and the notorious unreliability of estimates of Warsaw Pact defense spending.

To make matters worse, NATO's traditional technological lead over the Warsaw Pact is diminishing rapidly. Caspar Weinberger, the former United States Secretary of Defense, stated in 1986 that in a decade the Soviet Union had made up half of the West's technological advantage in practically all fields and that the Western lead, formerly estimated at ten to twelve years, had fallen to between four and six years.[3] Pentagon assessments show a substantial but decreasing technological superiority in available military technology, and a less marked but still decreasing superiority in deployed military technology.[4] In other words, the NATO nations, with their relatively strong innovative capacities and higher levels of defense spending, are not only being substantially out-produced by the Warsaw Pact nations but are also gradually losing their crucial advantage in military technology.

Table 1

Production of Selected Military Items in NATO and the Warsaw Pact in 1986*

Equipment Type	Warsaw Pact	NATO	Warsaw Pact/ NATO ratio
1. **Ground Forces**			
Tanks	3,700	1,000	3.7:1
Other Armored Fighting Vehicles	4,200	2,400	1.75:1
Towed Field Artillery	1,450	225	6.44:1
Self-Propelled Field Artillery	1,500	225	6.66:1
Multiple Rocket Launchers	575	500	1.15:1
Self-Propelled AA Artillery	150	0	—
Towed AA Artillery	200	0	—
2. **Aircraft**			
Bombers	50	25	2:1
Fighters and Fighter/Bombers	650	550	1.18:1
Transports	200	300	0.66:1
ASW	5	5	1:1
Military Helicopters	500	350	1.42:1
Utility/Trainers	55	250	0.22:1
3. **Naval Vessels**			
Submarines	8	7	1.14:1
Major Combatants	9	11	0.81:1
Minor Combatants	50	25	2:1
Auxiliaries	6	5	1.2:1

* Figures exclude France and Spain.
Source: *Soviet Military Power 1987*. US Department of Defense p.122.

Table 2

Production of Selected Military Items in NATO and the Warsaw Pact, 1975-84

Item	Warsaw Pact	NATO	Warsaw Pact/ NATO ratio
Tanks	25,500	10,990	2.3:1
Armored Vehicles	54,600	18,230	3:1
Mortars, Field Artillery + Rocket Launchers	24,500	4,500	5.4:1
Tactical Combat Aircraft	9,400	6,730	1.4:1
Attack Submarines	65	61	1:1
Major Surface Vessels	107	192	0.56:1

Source: Thomas A. Callaghan *The Future of Transatlantic Armaments Cooperation within NATO*. Paper presented at the 6th Armed Forces Communications and Electronics Association symposium, Brussels 23-25 October 1985.

There are several reasons for this sorry state of affairs. One factor, to be examined in more detail later, is the Warsaw Pact's program for acquiring Western technology. The primary reasons, however, are the Warsaw Pact's central planning which accords a high priority to defense and an evolutionary rather than a revolutionary procurement philosophy.[5] The Soviet Union is overwhelmingly the principal Warsaw Pact weapons producer, a fact that automatically allows long production runs with all the associated benefits of economies of scale and standardization. The high priority accorded to defense procurement ensures that the defense sector receives the best facilities and intellectual resources available and that new technology is developed—or acquired—and used as necessary. An evolutionary procurement structure also allows even "new" weapons to incorporate tried and tested—and possibly incrementally improved—technology while reducing total research and development costs and yielding benefits in areas such as ease of maintenance, maintenance training, reliability, interoperability, and logistics. Through this approach, the Warsaw Pact's aggregate military capability combines numbers and technology superior to NATO's even though item-for-item comparisons need not favor the Warsaw Pact.

NATO Procurement

This superiority owes as much to NATO's procurement weaknesses as to the Warsaw Pact's procurement strengths. NATO, as an Alliance of 16 sovereign, democratic nations, cannot impose a centrally planned procurement strategy upon its members. The United States does not dominate NATO procurement to the same degree as the Soviet Union dominates in the Warsaw Pact. While the Soviet Union accounts for over 90 percent of Warsaw Pact defense spending, the United States accounts for only about 60 percent of NATO's.[6] Indeed, for many NATO nations defense procurement is sufficiently substantial to be a major factor in their economies. Consequently, national defense procurement priorities and practices interact with national economic and industrial priorities and practices. Considerations such as employment, balance of payments, and the maintenance of national industrial and technological bases inevitably feature in the formulation of defense procurement strategies. Put another way, what may make sense specifically for the defense of the Alliance need not make sense for a nation's defense, economic, and industrial policies.

Duplicative Procurement and Structural Disarmament

The outcome of all this is a multiplicity of NATO weapons programs of staggering proportions. For most major weapons systems types, several NATO nations have their own domestic development and production efforts, or a variety of national and cooperative programs is in progress, as illustrated in Table,3. Even granting that strategic prudence may necessitate separate production lines in different nations, triple or quadruple efforts are difficult to justify. From an Alliance-wide perspective, such duplication is unnecessarily expensive and represents an inefficiency the Alliance can no longer afford. As Lord Carrington, NATO's former Secretary General, has stated "We quite simply cannot afford to perpetuate a system which, to quote only one example, has resulted in three main battle tanks—four if you include the Americans—being lined up to fight the same battle in the same place on the same day; and not even being able to use the same ammunition".[7]

Furthermore, the existence of many different systems has led to poor standardization and a lack of interoperability which prohibits

Table 3

Examples of Duplicated Procurement within NATO

System Type	No. of nations developing & manufacturing	Total type being developed and/or produced	Typical output per type			
			F	UK	FRG	USA
Combat Aircraft	5	14	200-400	200-400	175-325	1,000-2,600
Destroyers-Frigates	8	16	2-13	8-25	4-6	30-46
Main Battle Tanks	4	4	2,000	700-800	1,800-3,600	7,251
Anti-tank Weapons	7	11				
Ground-to-air Weapons	7	18				
Air-to-ground Weapons	7	16				
Air-to-air Weapons	6	8				
Ship-to-ship Weapons	7	10				

Sources: Brian Field ''Economic and Defense Resources: The Prospect'' *NATO Review* October 1985 pp.24-29
David M. Abshire ''Arms Cooperation in NATO'' *Armed Forces Journal International* December 1985 pp.66-70
Robin Beard ''New NATO Approaches to Improved Armaments Cooperation'' *NATO's Sixteen Nations* November 1986 pp.26-32

rationalization of logistics and maintenance so that NATO must have parallel support systems to operate and maintain its ever more sophisticated weapons systems.

Although duplicative procurement has, of course, never been ideal, its problems are compounded by the growing technological sophistication of defense equipment. This sophistication drives up unit costs and requires a large investment in research and development, which constitutes a major program overhead. Development costs of a combat aircraft, for instance, typically represent up to 25 percent of total program costs.[8] As a result, nations are compelled to accommodate rising equipment costs within relatively inflexible budgets and therefore frequently stretch procurement, an action which, in the long term only makes matters worse because operating production lines at below optimal levels causes unit costs to escalate still further.

This situation leads to what has been described as "structural disarmament". Even the United States is not immune, despite being better placed to enjoy economies of scale from much longer production runs. For instance, in 1982 eight of 13 production lines building US Navy aircraft were turning out fewer than 20 planes per year; three of these lines were expected to produce only six aircraft, causing "exorbitant unit costs and wasted resources". Furthermore, the number of production lines was set to increase such that unit costs would be still higher and, at existing rates of procurement and modernization, the Navy would enter the 1990s procuring 1960s technology aircraft. Other examples, from 1983, included the "stretched" procurement of the M-1 tank which was expected to add about $473 million to the total program cost, and the TOW II missile which was not scheduled to reach the most economically efficient production rate.[9]

Table 4 illustrates some more recent examples where unit costs have been increased by stretching procurement to remain within budgetary limits.

An additional strain on defense budgets arises from consistent over-optimism in forecasts of equipment costs and performance. For instance, an analysis of 80 US programs between 1950 and 1980 showed that only about 10 percent were completed within the original cost estimate and that raising estimates by 30 percent would have only given an even chance of achieving the cost forecast. To have a 90

Table 4

Decreases in Production and Related Increases in Unit Price

Item	Fiscal Year	Quantity	Unit Price	Increase per Item
MI Tank	1987	800	$2.3m	
	1988	600	$2.6m	$300,000
Blackhawk	1987	82	$5.2m	$1.3m
Helicopter	1988	61	$6.5m	
AV-8B Aircraft	1987	42	$16m	
	1988	32	$20m	$4m
EA-6B EW	1987	12	$36m	
Aircraft	1988	6	$59.6m	$23.6m*
E-2C Aircraft	1987	10	$44m	
	1988	6	$66m	$22m
Sidewinder	1987	627	$57,000	
Missile	1988	288	$150,000	$93,000

*Increase also due to upgrading.
Source: George C. Wilson "The Pentagon Diet: Bend and Stretchout" *The Washington Post Weekly* 26 January 1987 p.31

percent chance of completion within forecast costs, estimates would have needed to be 140 percent higher. Furthermore, only 15 percent of programs were completed on time and 30 percent did not achieve expected performance.[10] The United States is by no means unique in this: examples of actual costs exceeding forecasts are unexceptional for high technology weapons.

Operational Costs

In procuring new weaponry, performance and technical quality have been NATO's goals. Successive generations of military equipment essentially travel further and faster than their predecessors and are more accurate. These improvements have, of course, been facilitated by considerable increases in technical sophistication. But this sophistication can raise operational costs as well as acquisition costs, placing still more pressure on already stretched defense budgets.

All too frequently increasing complexity leads to decreasing reliability, while repairs, support, and maintenance necessitate large numbers of qualified technicians, advanced test equipment and high volumes of spare parts. As Lt.Gen. Leo Marquez, the United States Air Force deputy for logistics and engineering has observed "In recent years, we've been building airplanes and simply assuming away the support problem. And you people in the design community, you've come up with weapons that depend on a full support structure—including well-stocked spare parts depots, automated test equipment to diagnose failure in aircraft subsystems, elaborately equipped repair shops and all the people, cots, C-rations, and laundry facilities to run them".[11] This observation is borne out by a few facts and figures.

— An F-16 costs about $18 million, and operations and support over its lifetime total about $55 million; the US decision to purchase 216 F-16s during 1987 at a cost of almost $4 billion therefore carries an additional long-term budgetary commitment of almost $12 billion.
— On average, each fighter in the United States Air Force requires 17 maintenance specialists, or 408 per 24 aircraft squadron.
— Approximately 134,000 Air Force personnel, one third of all those enlisted, are involved in aircraft maintenance.
— The Air Force logistics system handles 890,000 types of spare parts with a total value of $40 billion.

It is also interesting to note some specific examples of the support problems encountered by the United States Air Force.

— The radio on the F-4 Phantom is located under the rear seat. This radio has a mean time between failures (MTBF) of between two and three hours so, on average, it must be removed for maintenance after every other flight. Before removal, however, a separate set of specialists has to disarm the seat ejection mechanism and then re-arm it after radio maintenance is complete.[12]
— The Mark II avionics system developed in the late 1960s for the F-111D was expected to achieve better than 60 hours MTBF

and maintenance man hours per sortie (MMHS) were expected to be about 1.4 hours. By 1980, MTBF was less than three hours and MMHS was 33.6 hours.[13]

These are by no means isolated examples.

It is encouraging, however, and not surprising that in the United States Air Force at least, the support problem is receiving particular attention and aircraft such as the F-18 have been designed with reliability and ease of maintenance very much in mind.[14] Such considerations are also written into the specifications of the Advanced Tactical Fighter (ATF) due to enter service in the mid-1990s to the extent that deployment of a squadron of ATFs should require "less than six" C-141 transport aircraft compared with 18 for an F-15 squadron and 24 for an F-4 squadron.[15] The Air Force has also sought to assign reliability and maintainability the same importance as cost, schedule, and performance as new systems are brought into the inventory.[16]

The United States Air Forces' emphasis on reliability and maintainability is to be welcomed, but the problems of expensive and time-consuming maintenance and repairs are still widespread throughout NATO's armed forces and every effort should be made to minimize them for future equipment procurements. The need to do so is certainly clear and more urgent than ever. As new technology improves weapons capabilities, the pace of combat and engagement distances increase. To capitalize fully on these characteristics, weapons systems must be integrated with data collection and processing facilities, and command, control, and communication (C3) systems. In other words, military operations will rely increasingly on weapons of growing sophistication working cooperatively with complex sensing and battle management systems. For instance, it is possible to envisage targeting data on a hostile surface vessel being obtained by reconnaissance satellite, transmitted through a communications satellite, processed by a command center, re-transmitted via satellite to the appropriate naval command center, processed again and transmitted via satellite to a friendly surface vessel and transferred to an anti-ship missile which would navigate to its target with the aid of a satellite navigation system. Analogous examples exist for air and land operations.

In a NATO context such processes could be rendered even more complex by including a series of electronic "black boxes" needed to render C3 systems compatible.

The success of any such operation obviously depends upon the successful functioning of each link in the network despite an adversary's attempts to disrupt them. Consequently, reliability is at a premium:

> Glamorous or not, it makes no sense to purchase highly capable weapon systems if the commander in the field cannot rely on those systems because maintenance requirements are too complicated, too frequent, too expensive, or the required spares or support equipment have not been bought, the supporting lifelines are operating as they must, or the required maintenance personnel are not available.[17]

Adding to these difficulties, the technological complexity and the expense of modern weaponry also often result in a paucity of "live" tests, a heavy dependence on simulation during development and operational training, and the deployment of hardware before development is complete.[18] These can lead to 10-fold disparities between test and operational performance.[19]

Defense Cooperation and Competition: Partial Solutions

The problems of duplicative procurement and the growing technological sophistication of weapons systems are at last being seriously addressed by the NATO nations, and increasing support for defense cooperation has been one of the primary results. Many instances of successful cooperation already exist and rising costs have heightened the need for further cooperative ventures.

Despite recent progress, however, the situation remains far from ideal and ample scope remains for both increasing the number of cooperative projects and improving the nature of cooperation. The number of cooperative ventures clearly could be increased because duplication on a large scale still prevails, although in practice there is no obvious mechanism for the wholesale reduction of programs already in progress. Nor would it be easy to bring the absolute costs of cooperative projects into line with purely national ventures with similar production runs. Cooperative ventures can carry an extra 30 percent in R&D costs and extra production costs of 10 percent.[20] As Sir George Edwards, former Chairman of the British Aircraft Corporation once commented: "The beauty of two countries cooperating on a weapons project and sharing everything 50-50, is that it only costs each two thirds".[21]

Coproduction is also less than ideal viewed from an Alliance rather than a national perspective. The F-16 European Coproduction Program, for instance, is acknowledged as an excellent example of successful international cooperation, but the F-16s acquired by Belgium, Denmark, the Netherlands and Norway cost an average of 35 per cent more than would have their direct purchase from the American production line.[22]

Despite their drawbacks, cooperative programs do enable nations to procure defense equipment more cheaply than would be possible on a purely national basis where R&D costs would have to be amortized over a shorter production run, and more acceptably—in terms of balance of payments and employment—than buying off-the-shelf from abroad.

Defense cooperation in its various forms, however, is not the only means of addressing NATO's defense production difficulties: increased defense competition and procurement divisions of labor also have a role to play. National armaments suppliers are often insulated from international competition, and opening national defense markets would undoubtedly drive down unit costs. According to one estimate, unit savings of 20 to 30 percent could be realized if NATO were to become a free trade area in weapons procurement.[23] This is a remote prospect since such a fundamental change would inevitably create major political, economic, and industrial upheaval. Even the nations which would probably emerge as net beneficiaries might experience painful economic dislocation because at least some of their producers would disappear. As for the net losers, the consequences need not be dwelt upon. Even so, scope exists for opening national defense markets to foreign competition beyond the present level.

Technology Transfer

One of the reasons why the Soviet Union is catching up in military technology is its large-scale, systematic program for obtaining Western technology. Using methods including illegal diversions of technology through "front" companies, espionage, and exploitation of open sources, the Soviet Union has acquired Western technology that has considerably enhanced Soviet weapons procurement and weapons quality. Naturally, Western nations have attempted to curtail this flow of technology but their preventative measures have had the unfortunate side effect of inhibiting technology transfers among the Alliance nations.

The forum for formulating and implementing Alliance export controls is the Paris-based Coordinating Committee on Multilateral Export Controls, COCOM. This body, which includes Japan and all NATO nations except Iceland, is used to establish common, national technology export regulations and also acts as a "clearinghouse" for exemption requests submitted by member nations. Essentially COCOM has three embargo lists: a munitions list, a nuclear list, and an international list. Munitions and nuclear exports, as to be expected, are relatively unproblematic. The international list, however, which covers "dual-use" items with potential military as well as civilian applications, is the subject of considerable debate. This list operates on a "phased" procedure. Less sensitive technologies on the list can be exported at national discretion. More sensitive technologies (or large exports of less sensitive ones) can also be exported at national discretion but COCOM must be notified. The most sensitive technologies require COCOM approval for export. COCOM operates on the principles of unanimity and anonymity so that any COCOM nation can veto an export license request.

The COCOM framework has received political and industrial criticism for various reasons, including the breadth of the international list, differing interpretation of exports allowable at national discretion, and delays in processing export licences. Evidence exists to support all these complaints. Fortunately, governments have addressed them and a steady improvement in the COCOM process has occurred, accompanied by a decline in misleading rhetoric. While room for improvement still exists, complaints now focus more on United States' export controls. These arise because the United States takes a more restrictive approach to technology exports than do its Allies, with the result that the Allies have had difficulty in obtaining American technology. This occurs not because the United States seeks to embargo technology exports to its Allies, but because an American condition of export is that subsequent re-exports are tracked and comply with American export regulations. Since these "end-user" controls apply to American components as well as finished goods, compliance causes an administrative burden and raises objections about the "extra-territorial" application of American law.

These problems have been eloquently described by Dr. W. Dekker, former President of the Philips Corporation. Dr. Dekker has pointed out that the application of American export regulations beyond the United States' borders offends "the ideals of national sovereignity" and adds "massively time-consuming and costly burdens" to conducting business. He explained that Philips might have to transport a component on America's control list across five European borders during the production, assembly and marketing process, and that in addition to each European country's licensing requirements, American approval was needed for each border crossing. Such approval was even required if the controlled components were produced by a Philips subsidiary in the United States. According to Dr. Dekker, this situation could have serious consequences:

> "European companies could be driven to find their component suppliers elsewhere or in other ways - and not necessarily in ventures with American companies on European soil, since they still fall under American unilateral jurisdiction. Both American and European companies are becoming increasingly alarmed at this state of affairs. And the risk is that, indeed, Western countries will grow apart because of practices that produce unintended outcomes".[24]

These sentiments are supported by a study of the effects of United States export controls conducted by American officials in the Federal Republic of Germany. This study covered 35 companies mainly in the electronics sector and found that:

— Several leading electronic equipment manufacturers said that they had begun to redesign their products to incorporate fewer US components... Where suitable components were also freely available from non-US suppliers, the companies were increasingly substituting them for US parts.

— A switch away from US products is said to be widespread among distributors of electronic components, mostly for lower value items. One distributor estimated that, as a direct result of its customers' concern about the controls, 20 percent of its sales were now being supplied from Japanese instead of US sources.

— Complying with the controls was said by US companies to add

> about 25 percent on average to their normal administrative overheads. The increase was relatively higher for smaller companies which generally found the controls more burdensome than did big groups.[25]

Essentially, the difficulties of complying with American export regulations provide an impetus to seek independence from United States technology so that, to some extent, other suppliers find it worthwhile to reinvent American wheels and Allies can be cautious about technology cooperation with the United States—particularly in sensitive areas—because of the difficulty in obtaining relevant information and components.

The United States has become more sensitive to this problem and has greatly streamlined its procedures for dealing with technology exports, and COCOM too has taken substantial steps towards harmonizing controls and enforcement procedures. Even so, remaining differences still provide potential for wasteful duplication of effort and further political acrimony. For example, the offer of formal American participation in the European Fighter Aircraft (EFA) was rejected even though the United States has much to offer in relevant technological expertise. This rejection no doubt stemmed in part from European concern about possible technology transfer problems. Indeed, bidding rules for EFA contracts insist on freedom to export all components. Bidders are warned that they must guarantee the free export of equipment supplied, or list in advance the countries to which export is prohibited. Since American or European contractors whose products incorporate American technology must comply with United States export regulations, their bids will be handicapped. Although export restrictions may not be overriding in awarding EFA contracts, past experiences where transatlantic defense cooperation has been hindered by technology transfer practices, the need to obtain American approval for moving components around Europe during manufacture and assembly, and the possibility of requiring American permission to export the finished aircraft all mitigate against contractors with American connections. As a result, such contractors are at a disadvantage when seeking business in the $30 billion program, and the project may be more costly than necessary if certain contractors are rejected through their need to comply with American export regulations.[26]

The Two-Way Street

The reservoir of transatlantic ill-will generated by problems with technology transfer was one of the factors behind Europe's unenthusiastic response to American proposals to explore a range of weaponry based on promising emerging technologies (ETs). Another factor was European concern about the potential American domination of the European defense market. The ratio of American defense sales to the European Allies, and of Allies' sales to America has been the subject of intense debate and it was feared in Europe that movement towards ET weaponry would result in a greater imbalance of transatlantic arms sales. Subsequently, the controversy surrounding the "two way street" declined somewhat as European manufacturers made inroads into the American defense market so that the value of American arms sales to Europe now probably is around only three times the value of European sales to the United States.

In any event, a relatively unified approach to ETs was adopted. Scope for intra-Alliance disagreements on ETs remains, however, since the United States is much better placed than Europe to make broad and rapid progress on the new technologies with potentially profound military implications.

The uneven traffic on the "two-way street" is, of course, no accident. The United States spends about three times more on military R&D than do the European Allies, and this disparity is compounded by Europe's fragmented, duplicative and uncoordinated efforts. Europe's shaky technological base renders it a poor competitor for the United States and also reduces its potential for negotiating joint transatlantic projects. Paradoxically, for Europe to cooperate with the United States, it must have a demonstrable capacity and willingness to compete with the United States. The way to achieve this is obvious but not simple: Europe must spend more on military R&D and, since it cannot hope to match American funding levels, it must get better value from its military R&D investments. This, of course, is easier said than done.

Conclusion

All these circumstances—duplicative procurement, poor reliability, conflicting national and Alliance priorities, technology transfer, and differing transatlantic capacities for exploiting new technology—have led to inefficient weapons production. As the level of technology embedded in weapons systems rises, these problems will only get worse unless corrective action is taken. The need for such action is evident: Warsaw Pact forces are catching up in technology and moving further ahead in numbers. Whether NATO can render the military balance more favorable is by no means assured and perhaps the best that can be achieved is a more gradual worsening of the situation. Even that, however, will be unattainable unless NATO exploits its resources more efficiently. So what are the technological resources at NATO's disposal and what should NATO do to take full advantage of them?

2

ELECTRONICS TECHNOLOGY

Since the Second World War, the electronics revolution has undoubtedly changed the face of modern military technology more than any other technological advance. Progress in other fields such as propulsion, explosives, and materials has, of course, been significant but no other technology has had such a radical and extensive effect on the design and employment of modern weaponry.

Electronics are all-pervasive in modern weapons systems. All but the simplest weapons such as small arms embody some form of electronics, and electronics are the basis of force multipliers and "enablers" such as communications, navigation aids, data processing systems, and a host of surveillance and reconnaissance systems. Electronics also contribute to the design and manufacture processes, to the testing of weapons systems, and to simulation and training.

This chapter describes the key trends in electronics technology and the issues raised by the widespread and increasing incorporation of electronics systems into Western force structures. Specific applications of electronics are mentioned for illustrative purposes only, with subsequent chapters describing in more detail the role that electronics play in current and planned weapons systems.

Growth in Military Electronics

As electronics have played an increasing role in the design, manufacture, operation, and integration of weapons systems, the proportion of defense budgets allocated to electronics naturally has risen accordingly. In World War One, for example, electronics accounted for only one percent of the American defense equipment budget, rising to six percent at the end of World War Two. In the 1970s, however, it rose to 20 percent and to almost 26 percent in 1986.[27] Associated with these increases have been commensurate rises in the number of armed forces personnel and defense contractors in electronics related fields. In the United States almost one third of armed forces personnel are employed in electronics related areas, and of the 100 largest defense contractors, 21 are electronics related.[28]

Furthermore, the contribution of electronics to modern weapons systems is set to increase. Again in the United States, almost half the budget for research, development, testing and engineering (RDT&E) for 1987 was allocated to electronics. The proportion is expected to rise to 54 percent by 1996, by which time between 35 and 40 percent of the procurement budget is expected to be electronics related.[29]

The electronics sector is characterized by an extremely rapid rate of technological innovation and obsolescence. Indeed, electronics innovations are so rapid that "upgrades" in weapons electronics are commonplace and the electronics incorporated into successive generations of weaponry change dramatically. In aircraft, for example, the days when a radio and possibly a radar gun sight represented an entire avionics suite are long gone. A "state-of-the-art" fighter today incorporates electronic flight and engine controls, radar, electronic countermeasures such as jamming equipment, communications, navigation and target acquisition systems, and a variety of weapons which rely heavily on sophisticated electronic hardware. In fact, as much as 30 percent of an aircraft's procurement costs can be electronics related. Similar situations surround weapons for use on land and at sea.[30]

As well as enhancing the capabilities of individual weapons, electronics have enabled a vast array of force multiplying systems to be developed. Radar and other surveillance and target acquisition systems, data processing equipment, and communications and navigation systems can now act in concert to ensure that weapons systems perform coordinated operations efficiently and effectively.

Further removed from actual combat, electronic systems also play increasingly significant roles in the design and production of weapons systems. Computers are used, for instance, as design tools for the full spectrum of military equipment, and manufacturing processes benefit from electronics-based systems such as machine tools and automated production equipment. Electronic simulation also aids the development process and is used increasingly for training and testing purposes. Indeed, such systems can be indispensible since the costs of some weapons prohibit sufficient "live" training.

While electronic systems have unquestionably generated remarkable improvements in weapons performance, they have also highlighted deficiencies in the procurement process by providing some of the best examples of procurement problems.

As noted earlier, costs tend to rise as weapons sophistication increases. This problem is particularly acute in electronics and is exacerbated by consistent under-estimates of development costs. The annals of procurement history are littered with examples such as the British Nimrod Airborne Early Warning System and the American DIVAD air defense system, where programs have been cancelled due to unexpected cost escalation. It is the exception rather than the rule for sophisticated electronic systems to be delivered on time and within original cost estimates. The same holds true for the reliability of electronics systems: until very recently the mean time between failures (MTBF) for electronics systems was almost invariably substantially below target.

Compounding these procurement difficulties is the need to deploy additional electronics capabilities in response to a continuously evolving threat. New, more capable hostile radars necessitate, for example, new, more capable countermeasures. This combination is extremely expensive, causing, for instance, up to 85 percent of an aircraft's lifetime cost to be accounted for by electronics.[31]

To make matters worse, electronics systems procurement is duplicated at national and international levels. The United States alone operates about 60 types of airborne radar warning devices, jammers and chaff/flare dispensers; and NATO aircraft operate 16 different types of radar warning devices.[32] Communications and data processing systems are also duplicated and, in many cases, are not interoperable.

These problems have at least been recognized and national and NATO authorities are seeking to address them. In the United States, Congress noted that "numerous unrelated efforts are under way . . . with little apparent attention being paid to achieving commonality" and, as a result, reduced 1985 funding for electronic warfare systems because of "the lack of a Department of Defense coordinated plan for electronic warfare development and production plans".[33] Increasing efforts are being devoted within NATO to producing common standards for a range of electronic equipment, particularly communications systems, and common procurement is receiving greater attention. Even so, many recently introduced national systems can only be rendered interoperable through the use of additional "black boxes" and, to cite just one example, truly interoperable Battlefield Information Collection and Exploitation Systems are unlikely to be in widespread use until 2000.[34]

The need to come to grips with procurement problems particularly in the electronics sector is paramount because, in the future, weapons performance and the successful conduct of military operations will depend even more upon electronics systems.

Technological Trends

Systems Integration

A key trend in modern warfare is increasing integration of geographically remote systems. As the complexity and pace of modern warfare increase, operations will require ever more capable systems to monitor and target hostile forces, data processing systems to appraise this information, and communications systems to disseminate it appropriately. Combat forces will also be more heavily dependent upon electronic warfare and electronic countermeasures in transit to their targets and during engagements.

For example, hostile aircraft could be detected by NATO's Airborne Warning and Control System (AWACS) aircraft which would then relay location data to a command position which would also be receiving other similar information from other surveillance assets. This information would then have to be processed and distributed through other command centers, to alert the necessary defensive aircraft and other air defense systems. All this would have to take place in the face of direct attacks on the various command "nodes" and despite efforts to deceive surveillance systems and disrupt communications using a variety of electronic warfare techniques.

Systems are now also in advanced development or in the early stages of production to mount similar operations against mobile land targets. Clearly, these will have to deal with larger quantities of data, more complex processing and distribution, and an increasingly potent electronic warfare environment.

Enhanced Weapons Capabilities

The trend towards growing sophistication in electronic systems used to support integrated operations is accompanied by increasing use of electronic systems on weapons. New weapons clearly will embody more versatile and more extensive electronics and increasing emphasis is placed on enhancing the performance and reliability of existing systems through sucessive electronics up-grades. In addition, electronic warfare (EW) systems will become more pervasive, causing a commensurate increase in EW resistant systems.

Equipment to locate, eavesdrop on or jam communications systems is now so commonplace that new systems make increasing use of frequency hopping and other defensive measures. Defensive avionics in particular will become more widely used and more capable. In the Central Region alone the Warsaw Pact operates about 10,000 radar-directed anti-aircraft guns, over 4,000 radar- and electro-optically-guided surface-to-air missiles (SAMs) and over 7,000 infrared-guided missiles.[35] At present, however, about 40 percent of NATO aircraft have either no electronic warfare systems or only a radar warning receiver. Despite the expense of radar jammers and other electronic countermeasures (ECM)—perhaps 10 to 15 percent of an aircraft's procurement costs—investment in them can yield substantial returns. At the beginning of the Vietnam War, for instance, only one or two radar-guided missiles had to be launched to down a single aircraft but ECM equipment eventually increased the number to over 60.[36] Infrared countermeasures are also necessary, however: of the 140 aircraft lost in combat between 1975 and 1985, over 90 percent fell victim to infrared guided missiles.[37]

Sensor technology will also continue to increase in capability as weapons systems incorporate more powerful electronics to improve target location and discrimination. No fundamentally new sensor techniques are likely to appear since the electromagnetic spectrum and atmospheric conditions impose physical limits on the generic techniques available and all possibilities are known. Consequently, it is only the emphasis on particular sections of the electromagnetic spectrum may change. Millimeter waves, for instance, are likely to be exploited more extensively as new technology—notably in gallium arsenide chips—facilitates use of this particularly promising field.

Millimeter waves can be used actively—like radar—or passively—like infrared—and in active mode especially can achieve high resolutions with modest size antenna and can produce a very narrow signal beam with little "leakage" for an opponent to detect. Weapons will also tend to incorporate several types of sensor as microcircuitry advances produce smaller sensors and processors. Such "multi-mode" sensors provide greater target discrimination and greater resistance to countermeasures. Another possible trend is increasing use of longer wavelengths in radars since these may be more effective against "stealth" vehicles.

Software Production

Demand for increasingly lengthy and complex software is rising sharply as electronics are used more extensively. The American Joint Surveillance Target Attack Radar System (JSTARS), an airborne ground surveillance system, will require 1,700,000 lines of computer code per aircraft, 700,000 lines of which are new. The British Experimental Aircraft Program (EAP), a technology demonstrator aircraft, incorporates 500,000 lines and aircraft incorporating up to 10,000,000 lines are anticipated.[38] Computers can themselves assist with software production but additional software tools are necessary as is some substantial software standardization. NATO has adapted a computer language known as "Ada" as a standard in order to rationalize software production and use, but progress is slow and software may never be completely uniform due to the peculiarities of computer languages, systems, and certain applications.

Software demand will certainly lead to increasing use of software "modules" for assembling into larger programs and will accelerate software standardization, design tools and test techniques.

Expert Systems

A key aid to software production will be the use of "expert systems" or artificial intelligence. Artificial intelligence—in the sense of autonomous decision making—is a distant prospect but "intelligent" or "expert" systems which can be used for specific, well-defined tasks are already available and are being used in many civilian and military fields. Existing military applications include equipment fault diagnosis and assistance with software production. Other near-term applications, probably better described as advanced automated systems, include automated tank cockpits, semi-autonomous battlefield robots, sentry systems, and gun loaders. Systems with more general applications based on limited abilities such as robot vision and speech recognition are also in advanced development.

Optical Fibers and Computers

The trend towards replacing copper cables with optical fibers for data transmission will undoubtedly accelerate. Optical fibers are already used to some extent, but enormous scope for further use remains.

Such fibers are typically one twentieth the diameter of the cables they replace, require no shielding to guard against emissions or outside interference and carry no risk of generating sparks. Furthermore, their transmission rates are higher and "repeaters" are needed at intervals of about 30 kilometers compared with two kilometers for standard coaxial cable. For battlefield cable communications, the use of fiber optics could cut total weight by 67 percent, reels by 60 percent, and transport vehicles by 80 percent.[39] Fiber optics could also replace much standard wiring in many weapons systems. Already a wire-guided missile using optical fiber is in development, and shipboard safety systems, optical gyroscopes, "stealth" avionics, aircraft flight control, and a host of other applications are also planned.

A development which ultimately may genuinely revolutionize electronics is the optical computer. Viable systems are not likely to appear before the end of the century but the potential is enormous. Such systems would process light rather than electrical signals and so would operate thousands of times faster than today's supercomputers. In addition, optical devices would be able to process simultaneous inputs and would be immune to outside interference. Provided they could be sufficiently miniaturized and costs were acceptable, optical computers would supplant electronic systems and undoubtedly would create new weapons and mission possibilities. For the next decade, however, the challenge is to establish the basic technologies rather than to seek applications.

VHSIC and MIMIC

Even without optical computers technological progress now offers the prospect of dramatic improvements in both performance and reliability. While such developments always seem to have been "just around the corner" they do now appear to be in sight although their impact is likely to be limited in the short term. Programs to develop very high speed integrated circuits (VHSIC) for digital systems and monolithic microwave integrated circuits (MIMIC) for analogue systems will, for instance, produce electronic systems with an unprecedented combination of high capacity, small size, flexibility, and reliability. The VHSIC program, for example, has an initial goal of producing chips with feature sizes as small as one micron, containing about 100,000 devices operating at 25 Megahertz and ultimately will achieve feature sizes of 0.5 microns, with about 300,000 devices operating at about 100 Megahertz. In contrast, commercial chips typically have feature sizes of two to three microns, contain about 10,000 devices, and operate at 8-12 Megahertz.[40] In practical terms, VHSIC will enable the production of electronic systems requiring fewer chips, less connective wiring, and fewer boards. VHSIC-based systems will also be "fault tolerant" with built-in test circuitry and the flexibility to perform several distinct tasks. These characteristics will translate into concrete operational benefits. Already, a VHSIC module is being produced to replace the MIL-STD-1750A aircraft mission computer. This will operate at three to four times the speed while occupying a third of the space.

VHSIC technology will also yield substantial savings in maintenance. If, for instance, the avionics on the F-16 A/B were VHSIC-based, the 20-year ownership costs would be cut by 47 per cent. For each F-16 the 437 types of spares for Line Replaceable Modules would be cut to 43, mean time between failures for each aircraft would rise from 7.3 hours to 35 hours, sortie rate would rise from 3.1 to 4.5 per day, flight line avionics maintenance personnel for each 72-aircraft wing would be cut from 142 to 71, and the 38-man avionics intermediate shop would be eliminated.[41]

No plans exist to retro-fit the F-16 with VHSIC systems but next-generation aircraft such as the Advanced Tactical Fighter (ATF) will be so equipped and are thus expected to achieve the combination of performance and reliability which existing aircraft have been unable to attain. The ATF force should have about 90 percent availability compared with 80 percent for the F-15, and should achieve this with half the maintenance personnel of the F-15. According to one commentator "VHSIC and other advanced microcircuits may finally put to rest the notion that there need be trade-offs between avionics systems performance and reliability, or that more complex and advanced electronic systems inevitably lead to lower reliability and maintainability".[42]

Conclusion

The increasing use of electronics in weapons and as force multipliers should give procurement authorities grave cause for concern. The propensity for prices to rise beyond predictions and for reliability to fall short of expectations will place extreme pressure on defense budgets. In this area more than any other, it is essential to eliminate duplication and to attach paramount importance to reliability and standardization as matters of urgency. Otherwise, the more extensive use of duplicated systems placing heavy demands on maintenance resources is clearly a recipe for long-term problems.

Technological advances such as VHSIC plus existing political pressure to rationalize procurement will undoubtedly but only gradually reduce the burdens stemming from poor reliability, heavy maintenance requirements, and inefficient duplication of effort.

Policy makers should seek to accelerate the resolution of these difficulties by ensuring that standardization and the elimination of duplication are accorded the priority they deserve, and that weapons technologists assign the same priority to reliability and ease of maintenance as they do to "traditional" performance parameters. Otherwise with rising budgetary pressures and maintenance burdens electronics technologies will become NATO's Achilles heel instead of its great defense asset.

3

BATTELFIELD TECHNOLOGY

The application of new technology continously improves weapons capabilities generally with little effect on overall mission concepts. Technologies now in development or in the early stages of procurement, however, may cause a substantial change in concepts of battlefield operations.

Technological progress, largely in the electronics field, now offers the prospect of substantial advances in three areas: in surveillance and target acquisition; in command, control and communications, and data processing; and in delivery systems and munitions. Advances in any of these areas have obvious military utility, but when combined they may offer radical improvements in capability to the extent that traditional military missions may be conducted much more efficiently and in new ways and, indeed, new mission possibilities may be created. This section describes those technologies and then turns to the major operational and procurement issues associated with them.

Surveillance and Target Acquisition

Advances in the miniaturization of electronic components have been largely responsible for dramatic improvements in a wide variety of sensor technologies. Radar systems, for instance, can now track and even identify large numbers of targets at long ranges, and sophisticated radar, optical, infrared, or millimeter wave sensors can be fitted to munitions as small as a cannon shell.

The best example of a long-range battlefield surveillance system is the United States Joint Surveillance Target Attack Radar System (JSTARS), due to become operational in the early 1990s and intended specifically for use in Europe. JSTARS will use a sophisticated radar mounted on E-8A aircraft (modified Boeing 707s) to monitor activities on the ground up to almost 400 kilometers from the aircraft. Current plans call for procurement of 10 JSTARS aircraft capable of remaining airborne for up to 20 hours, plus 107 mobile ground stations consisting of a mixture of five-ton trucks and jeep-sized vehicles.[43] JSTARS will also relay data directly to attack aircraft equipped with suitable receivers. Communications will be via the Joint Tactical Information Distribution System (JTIDS) and the Surveillance and Control Data Link (SCDL). A network of navigation beacons will provide position data until the Global Positioning System (GPS) becomes operational. Thereafter, the beacons will serve as a back up.

Since Europe is already extensively mapped, terrain details will be fed into the system and superimposed on the radar picture so that JSTARS will not only be able to locate enemy vehicle formations but will also be able to project their likely progress in view of the terrain features they must negotiate. JSTARS will also be able to up-date missile target information while missiles are in flight although, as yet, no missile that can operate in this way is available.

In terms of vulnerability to hostile action, JSTARS is described as "survivable but not immortal" since it will operate in defended airspace, and even if withdrawn to 200 kilometers from the forward line of troops (FLOT) could still provide information of activity well into hostile areas. Susceptibility to jamming also appears very low. Jammers of sufficient power or accuracy to degrade JSTARS significantly would be difficult to construct and operate. JSTARS' highly directional radar already provides substantial resistance to jamming, but up-grades could improve resistance still further. Enemy jammers could also be attacked directly if necessary.[44]

JSTARS will provide enormously useful information to military commanders. By locating hostile forces as they move towards the FLOT, it will be possible to deploy defensive forces more efficiently, and targeting data on forces close to and well behind the FLOT will facilitate attacks on those forces.

JSTARS is not the only American system which will survey the battlefield and beyond. The American OV-ID "Mohawk" aircraft already in service carries a moving target indicator (the APS-94 sideways looking radar), an infrared mapper, and reconnaissance cameras. Its capacity and range is less than JSTARS and it must operate within range of several types of Soviet surface-to-air-missiles. The ASARS-II radar mounted on the TR-1 aircraft also provides a similar capability but, again, far less than JSTARS: area of coverage, accuracy of location, detection range, and resistance to electronic countermeasures are all inferior.[45]

Other Alliance nations also operate or are developing airborne battlefield surveillance systems. Although these other "stand-off" surveillance systems will not offer JSTARS' capacity, long-range coverage, and command capability, they will nevertheless provide extremely valuable information to local commanders. The United Kingdom is developing the Airborne Stand-Off Radar system, ASTOR-I (for moving targets) and ASTOR-C (for fixed and moving targets). ASTOR-I, based on an "Islander" aircraft, and ASTOR-C, based on "Canberra" aircraft, while undeniably useful for battlefield surveillance, will be substantially less capable than JSTARS. In addition they are intended to operate without "real time" data links, recording data for analysis after aircraft recovery, although experiments have been conducted to establish the feasibility of using real time links in conjunction with JSTARS ground stations.

France is also developing its Observatoire Radar Cohérent Héliporté d'Investigation des Eléments Ennemis (ORCHIDEE). Expected to enter service in 1995, ORCHIDEE will consist of a battlefield surveillance radar mounted on a "Super Puma" helicopter. This system will be able to detect and locate vehicle concentrations at a range of about 150 kilometers and will transmit surveillance data in real time using the French forces' RITA communications network. ASTOR, ORCHIDEE, and JSTARS were originally independent programs but in late 1987 it was decided to render them interoperable.[46]

Stand-off surveillance systems are not the only means of obtaining real time information about hostile ground forces beyond the FLOT. These systems are increasingly being complemented by penetrating unmanned aerial vehicles (UAVs) which, as well as locating and identifying hostile forces, will soon be able to perform tasks such as target designation and jamming. Several NATO nations already operate UAVs to perform reconnaissance missions, providing real time images using optical or infrared techniques, or to take photographs which are available after vehicle recovery. Although these functions are the most common, some systems in development or in early production can perfom other tasks such as electronic intelligence gathering, surveillance by radar, laser target designation, communications relay, and direct attack of targets.

Existing UAVs come in many shapes and sizes, ranging from those which resemble small jet aircraft to those which are like tiny helicopters. The general trend, however, is towards small, propeller-driven vehicles resembling model aircraft which provide long endurance and small radar and infrared signatures. These relatively inexpensive vehicles can loiter in the vicinity of hostile forces since they are very difficult to detect and destroy, and can carry a variety of equipment such as sensors, target designators, and jammers which, thanks to the electronics revolution, are now small and powerful enough to be incorporated in such small vehicles.

UAVs come in two main types, remotely piloted vehicles (RPVs) which are controlled by operators, and drones which fly preprogrammed routes. The utility of such systems in combat was vividly illustrated by Israeli forces in Lebanon in 1982 when RPVs were used to locate Syrian surface-to-air missile (SAM) sites in the Bekaa Valley. Loitering at altitudes of up to 1.5 kilometers, Israeli RPVs radioed back extremely clear television pictures of the camouflaged sites and provided precise coordinates for air-to-ground missiles. This information enabled Israeli forces to destroy all 28 SAM sites with the loss of only one aircraft which fell victim to a shoulder-launched missile.[47]

One of the most ambitious UAV projects was the American Aquila. The Aquila program was cancelled in February 1988 in order to accommodate cuts in the defense budget and because American UAV programs are being rationalized to allow four basic types to meet all service needs for information gathering UAVs. An effort is under way therefore to transfer Aquila technology to the new programs which should produce operational systems by the end of 1991.[48] Aquila is about two meters long with a wingspan of almost four meters and is constructed largely from composite materials. A small piston engine drives a ducted pusher propeller and the vehicle's speed varies from 90 to 180 kilometers per hour. Maximum altitude is about 3.6 kilometers and endurance about three hours. Aquila is launched from the back of a flat-bed truck and is recovered by a net capture system. Aquila carries a television camera or infrared sensor and is equipped with a laser designator/rangefinder, all stabilized by an on-board microprocessor which can keep them locked on to their chosen target regardless of their or Aquila's motion. Images from sensors are transmitted to operators who can aim them and zoom in on areas of interest. Target locations can be established precisely for transmission to attack systems such as conventional artillery or the Multiple Launch Rocket System and, if desired, the laser designator can be used to illuminate targets for laser-guided systems such as the Copperhead artillery projectile and the Hellfire and Maverick missiles. Aquila can then monitor the attack. Control and data links are extremely jam resistant and Aquila's radar and infrared signatures are very low, so the system is highly survivable.[49]

Although some development difficulties such as sensor stabilization have been troublesome, UAVs derived from Aquila and other similar systems will undoubtedly be enormously useful on the battlefield in the future. They will provide up-to-date reconnaissance, target location, and damage assessment information and will perform target designation in areas which would be difficult or impossible for manned systems to enter.

As noted earlier, however, these are by no means the only applications of UAVs. Systems are being developed to gather electronic intelligence, to conduct electronic jamming, and to attack hostile forces directly. The United States, for instance, is developing the Brave 200 "fire-and-forget" armed drone to seek out and destroy enemy air defense radars. Fifteen of these drones will be launched from a standard-sized container carried on a truck or trailer. Each drone will follow a preprogrammed flight path and will loiter for up to eight hours until it detects a radar emitter which it will then attack. This system will obviously make life easier for aircraft penetrating hostile airspace since SAM operators will risk destruction when they switch on their radars.[50]

The United States is not alone in pursuing UAV technology. Including private-sector initiatives, almost 70 UAV types are in development or production in NATO nations and less than half are American.[51] The United Kingdom, for instance, is developing the Phoenix battlefield surveillance/target acquisition RPV equipped with radar and infrared sensors. Other programs such as the Falconet, Firebird and Raven will produce UAVs to conduct missions such as electronic countermeasures, reconnaissance, and target acquisition and to act as decoys. The Federal Republic of Germany also has comparable programs such as the Tucan, the KZO and the KDAR. Canada, France, Italy and Greece also are developing UAVs of various types.[52]

Airborne stand-off sensors and UAVs are by no means the only systems to provide battlefield surveillance and targeting information. Attention is also being paid to acoustic, seismic, magnetic, infrared, and millimeter wave sensors which can be lofted into enemy-held territory by artillery or aircraft, or left behind by retreating forces. Such sensors were used extensively in the Vietnam War, but electronics technology is now facilitating the development of similar systems which are both smaller—and therefore difficult to detect—and more sensitive.

More traditional systems are also benefitting from the application of new technology. Such systems include penetrating reconnaissance aircraft, ground-based surveillance radars, and a host of electronic emission monitors which can provide vast quantities of useful and not so useful data. Electronic navigation systems, for instance, improve the location data furnished by reconnaissance systems, and advances in radar and data processing are greatly enhancing the quality of reconnaissance information.

Communications and Data Processing

To exploit the data provided by all these systems effectively, it must be communicated reliably either directly to those who can use it or to command centers where it can be analyzed and then disseminated as appropriate. These vital functions require enormous communications and data processing capacities which must handle not just surveillance information but also command messages, location and status reports, logistic information, and any number of other signals needed to support military operations. Furthermore, all this must be executed despite adversaries' attempts to jam, deceive and destroy the systems performing these tasks.

At present, NATO communications systems and data processing equipment are overwhelmingly national prerogatives, so many different systems are in operation and development. In general, these systems are not directly interoperable though programs are in hand to introduce a measure of interoperability through interface systems and the development of common standards and protocols.

New communications systems, although not interoperable, employ similar techniques and architectures to meet present and future demands. To complicate the adversary's tasks of jamming, interception, and location, communications systems will increasingly use frequency hopping techniques whereby transmitters and receivers switch simultaneously to different frequencies very rapidly. The hops are pseudo-random and, depending on the system, vary from 150 to about 2,000 hops per second. Other techniques can be used such as signal compression, millimeter wave transmission (as opposed to HF, UHF and SHF) or even laser communications. But the general trend is towards frequency agility plus automatic electronic encryption and decryption. Particularly at low power levels, transmissions from such systems can be virtually indistinguishable from normal background radio noise.[53] To reduce the effects of attrition, communications systems can also be networked so that each transceiver acts as a relay station and the systems' built-in electronics wizardry routes communications as necessary.

Many communications systems are needed to meet various service and mission requirements but description of a few well illustrates current uses and trends.

The British Ptarmigan and the French Rita digital communications networks now in service provide tactical trunk communications and combine the addressability of the telephone with the mobility of the radio. They both employ digital, automatic switching and relay systems which shunt communications around damaged areas without any operator intervention. They also provide automatic encryption and decription and can handle speech, telex, facsimile, and computer communications. Each user—fixed or mobile—can be called simply by dialling his number and the systems' electronics find a route through.[54]

The American Joint Tactical Information Distribution System (JTIDS) acts as a combined communications and data processing system. Designed for combined operations, JTIDS will be purchased by several NATO nations. Providing a secure, jam resistant link between air, ground and naval forces, JTIDS is already deployed on a limited scale to provide, for instance, communications and data links between Airborne Warning and Control System aircraft and associated fighter aircraft, and will be deployed far more extensively in the 1990s. JTIDS information is broadcasted omnidirectionally to be received by any terminal within line of sight and each user can select messages according to need. A technique known as "time division multiple access" permits thousands of users to up-date and access information without mutual interference. The system can also be used passively to permit a user to receive information while maintaining radio silence. In active mode, however, user status—such as aircraft location, altitude, gound speed, direction, fuel and weapons reserves, and radar signature returns—will be fed into the JTIDS net. Consequently, a pilot, for example, could use JTIDS to furnish information on SAM locations, alternate recovery bases, loctions of hostile and friendly aircraft, and locations of hostile and friendly ground forces. He could also discover which targets are assinged to whom. Information beyond visual range—selectable between about eight to 480 kilometers—can be displayed on a cathode ray tube which depicts data using color-coded symbols.[55]

Another system to be fielded in the late 1980s is the American Precision Location Reporting System (PLRS) which will enable position data for all suitably equipped units to be ascertained with considerable precision. PLRS user terminals weighing about 10 kilograms will interact with each other to provide location data to within 15 meters for personnel, 25 meters for land vehicles, and about 50 meters for aircraft.

PLRS is a division level system which will consist of a master station (plus back-up) controlling up to 400 user units in an area of 300 kilometers square up to an altitude of 15,000 meters. About 2,000 users in a corps area will be able to operate PLRS on a non-interfering basis. Essentially using triangulation techniques, the system assembles position data on all users and provides a limited digital communications capability. The system is fully automatic and relieves conventional communications systems and operators of the substantial burden of reporting position. The system is interoperable with the satellite Global Positioning System and will enable, for instance, personnel, vehicles, and aircraft to navigate in darkness and bad weather and to keep track of each other. Commanders at all levels will, probably for the first time in history, know the precise location of their own forces during battle.[56]

Work is also in progress to produce a PLRS/JTIDS hybrid (PJH) system to enter service in 1990. PJH will facilitate data communication, position location reporting, navigation and identification of friendly forces. The system will, for instance, allow fire support requests to be distributed with the minimum of operator interference, reliable distribution of information about friendly aircraft locations to SAM units, and data and position information distribution for logistics support and integrated ground/air operations.[57]

Data processing will feature increasingly in operations at all levels as the complexity and pace of combat is increased by the deployment of ever more capable sensors, communications and weapons systems. For coordinating data on emitters, SAM sites, mobile forces, aircraft, logistics and so forth, and for distributing this data as needed, data processing equipment is essential.

Increasingly, data processing equipment will assimilate and distribute data automatically, and will form an integral part of communication systems. For illustrative purposes it is sufficient to describe just a few systems which essentially mark opposite ends of the data processing spectrum ranging from small systems used to manage artillery engagements up to enormously powerful systems used to provide a comprehensive appraisal of the battlefield.

An example of the former is the American TACFIRE (Tactical Fire Direction System) system, first deployed in 1967, which processes observers' artillery fire requests and allows a battalion fire support coordination center to forward them to the appropriate battery by presenting data on weapon status, ammunition stocks, and meteorological data. TACFIRE can handle tasks for up to 100 fire units while simultaneously storing data on over 1,300 targets. This obsolescent system is to be replaced with the AFATDS (Advanced Field Artillery Tactical Data System) which further automates the transmission of data from the observer to the battery and incorporates air support operations into fire support operations.[58]

At the central systems level, France, the Federal Republic of Germany, the United Kingdom, and the United States are each developing data processing systems to furnish comprehensive pictures of the battlefield. These systems—known respectively as SACRA, HEROS, WAVELL, and MCS—are also intended to become interoperable.[59]

WAVELL will allow all tactical command cells down to brigade level to obtain data on plans and orders, intelligence, current operations, terrain, order of battle, air defense cover, logistics, etc. Essentially, it will be an electronic "letterbox", accessed by Ptarmigan, which will allow users to exchange data.

By 2000 WAVELL and other similar systems, or their successors, are planned to be integrated into a Battlefield Information Collection and Exploitation System (BICES) which will enable users to obtain information on the entire European theater. A commander, for instance, will be able to call up the locations of all artillery groups, tank formations, or any selected element on a 50 kilometer segment of the front while, elsewhere, another commander views a 300 x 300 kilometer area portraying all SAM locations.[60]

Integrating systems to create BICES is far from straightforward, however. Nine West European countries are cooperating on the development of the European Data Distribution System (EDDS) which will perform some BICES functions but which might not be compatible with United States and Canadian data transfer systems. The United States operates its Limited Operational Capability Europe (LOCE) system to process intelligence data. LOCE interfaces directly with theater and tactical sensor systems, correlates their reports and assembles a comprehensive data base on hostile forces. How systems such as EDDS and LOCE will be enlarged and adapted to create BICES and whether BICES will be based on a large central processing facility or on a decentralized network of local data systems must still be decided.[61]

Smart Weapons

The ultimate reason for developing elaborate sensing, communications, and data processing systems is, of course, to facilitate the effective use of munitions systems to destroy and disrupt an adversary's military assets. These munitions systems are also benefitting from technological advances in various areas and some technologies under investigation offer the prospect of dramatic improvements.

The most obvious trend in battlefield munitions is towards increasing "smartness". Smartness is now being engineered into everything from cannon shells, mortar projectiles, and mines to anti-tank missiles, tanks and helicopters. While the "magic bullet" guaranteeing one kill per shot is still some way off, the time is disappearing when targets are destroyed by placing overwhelming explosive power in their general vicinity and kill rates are determined by the laws of probability.

One of the best examples of this trend is in artillery. In general, artillery fire accounts for about 60 percent of the casualties in battle and about one percent of tank kills, with each such kill requiring the delivery of about 250 rounds.[62] Terminally-guided rounds such as the American Copperhead, however, could reduce this figure to only one or two rounds per kill. Using this system, a forward observer or a UAV such as Aquila illuminates the target with a laser designater and the Copperhead projectile homes in on the laser spot. Similar techniques can be used to guide anti-tank weaponry launched by infantry and aircraft.

The drawbacks are the need to maintain a line of sight to the target and to carry special designating equipment. In addition, the observer is liable to detection and can be—at the very least—forced by suppressive fire to break his line of sight. Consequently, work is in progress to develop autonomous homing projectiles which need only be launched to the general vicinity of attacking armor to enable on-board sensors to locate and destroy their targets without human intervention. With this type of munition, an observer needs only a radio, a map and an awareness of his own position to call down fire. This greatly increases the number of potential observers and does not require them to expose themselves to suppressive fire.

The munitions capable of performing in this way, such as the American Sense and Destroy Armor (SADARM) should be available in the early 1990s. A 155 millimeter projectile could carry two SADARMs which would descend relatively slowly by parachute at the end of their flight, automatically scanning a circle of 75 meters radius with infrared and millimeter wave sensors. On detecting a target, the SADARM munition would explode producing an Explosively Formed Penetrator (EFP), also known as a Self Forging Fragment (SFF), essentially a slug of metal travelling at up to 3,000 meters per second, which would penetrate the target's top armor. A refinement of this technique is for the munition to remain active if no target is located and to act as a mine, laying in wait for a target to enter its lethal radius.[63]

The use of terminally guided sub-munitions (TGSMs) such as SADARM will add the role of anti-armor engagement to conventional artillery so that in the future, less emphasis may be placed on tanks to destroy enemy armor in order to procure more artillery.

Another trend in artillery is towards the greater use of rocket projectiles. France, the Federal Republic of Germany, Italy, the United Kingdom and the United States are deploying the Multiple Launch Rocket System (MLRS) which can lay down an extremely heavy salvo very rapidly. Each mobile MLRS launcher carries twelve 230-millimeter rockets which can be fired in rapid succession to ranges beyond 35 kilometers. The current warhead contains 644 dual-purpose conventional sub-munitions which are effective against personnel and lightly armored vehicles. Each twelve rocket salvo can therefore place 7,728 sub-munitions on a target area. On average the system places one bomblet per 20 square meters and the total effect is devastating against unhardened targets.[64]

A terminally guided warhead (TGW) is being developed jointly by France, the Federal Republic of Germany, the United Kingdom, and the United States to enable MLRS to counter armored targets. Each MLRS rocket will carry three TGWs which will glide over the target area after release from the carrier rocket, searching for armored vehicles using infrared and/or millimeter wave sensors. The search area is between 300 and 500 meters wide and three to five kilometers deep in the direction of travel. Another proposed variant would equip each rocket with six SADARM sub-munitions. The Federal Republic of Germany and Italy also intend to use MLRS to lay AT-2 mines which are effective against any vehicle. Each rocket could carry 28 of these devices.[65]

The United States is also working on the Army Tactical Missile System (ATACMS) which will use a standard MLRS launch vehicle to launch a single missile equipped with 16 to 24 terminally guided sub-munitions over a range of about 150 kilometers. The sensor used would be highly discriminating, able to attack critical areas and not just "hot spots". A single warhead version is under development and a nuclear-armed variant is a possibility as a follow-on to the lance missile.[66]

Other systems are also being developed to enable ground forces to engage armored vehicles autonomously at relatively short range.

France, the Federal Republic of Germany and the United Kingdom are, for instance, developing a "Third Generation Anti-Tank Missile" (TRIGAT). This will be produced in two versions, one for launch from ground vehicles and helicopters and the other as a man-portable type. The former will have a range of about five kilometers and will use an infrared guidance system which will give TRIGAT a "fire-and-forget" capability. Once launched, the missile will home in on its target automatically, enabling the user to disengage immediately. The Federal Republic of Germany and the United Kingdom are developing a mast-mounted installation for TRIGAT for use on specially adapted ground vehicles. This will elevate a four-round launcher and sighting system to a height of up to 12 meters so the vehicle will be able to engage targets from behind cover. The launcher will be reloaded automatically from a magazine in the vehicle. The man-portable version will have a range of about two kilometers and will require the operator to keep the target in an optical sight with a laser communications link providing guidance instructions to the missile. This version will be extremely agile and will have "soft" launch characteristics enabling it to be fired from inside a building. Both versions will have a tandem warhead capable of defeating reactive armor. The first charge detonates the reactive section allowing the following charge to penetrate the conventional armor beneath.[67]

A particularly interesting system is the fiber optic guided missile, under development in the United States, France and the Federal Republic of Germany. This type of missile has a nose-mounted television camera or other imaging device which transmits a picture via optical cable to an operator who guides the missile onto its target. Unlike traditional wire-guided missiles which require a line of sight between operator and target, and have a range of at most a few kilometers, fiber optic guided missiles require no line of sight and will have a range of up to 15 kilometers. Operators are thus further removed from the FLOT and the missile itself is relatively unsophisticated and cheap since the fiber optic cable has a data transfer capacity 5,000 times that of a metal cable so "intelligence" can be built into the control console rather than the missile.[68]

Several types of weapon based on the fiber optic guidance principle are under development. In France and the Federal Republic of Germany, anti-tank and anti-helicopter versions to be launched from truck-mounted firing units are under development. In the United States, a man-portable anti-armor version, the Army Advanced Anti-tank Weapon System Medium (AAWS-M) with a range of about two kilometers is under investigation as a possible replacement for the Dragon anti-tank weapon.[69] The Fiber Optic Guided Missile (FOG-M) is being developed as an anti-helicopter missile with a range in excess of 10 kilometers and this is expected to be followed by an anti-tank version with a similar range. These should enter service in 1992.[70]

Although the range of fiber optic guided missiles will be much shorter than that of autonomous projectiles, it is greater than other man-guided systems while retaining the "man in the loop", a situation preferred by some analysts who doubt the capabilities of autonomous systems.

Mines

Increasing attention is being paid to mines which are much more sophisticated than "dumb" bombs which must be driven over to detonate. Small, "off-route" mines which blast a projectile into vehicles passing nearby are already available and can be laid by hand, or mechanically or by aircraft, rockets and artillery. Using a variety of sensors, it is even possible to select particular types of vehicle for attack. One such system is the Anglo-French Ajax/Apilas which has a range of up to 200 meters against vehicles travelling at up to 80 kilometers per hour. The weapon can remain dormant for up to 60 days and is "awoken" by seismic and acoustic detectors which activate an infrared targeting system which can be programmed to select particular types of target. The Federal Republic of Germany, France, and the United Kingdom are jointly developing a more capable system known as the "aimed controlled effect anti-tank mine" (ACEATM), which will be even more capable.[71] Work is also in hand to produce "stochastic" mines which occasionally jump around to seek targets or which can be programmed to move into cleared areas.[72]

A novel concept for clearing minefields is the Catapult Launched Fuel-Air Explosive (CATFAE) due for deployment with the United States Marine Corps in 1991. The system consists of an armored chassis fitted with 21 fuel-air explosive rounds which can be catapulted ahead of the vehicle. These detonate on impact and create a very high overpressure which neutralizes anti-tank and anti-personnel mines. The full complement of rounds is sufficient to clear a path 20 meters wide and 300 meters long.[73]

Tanks and Artillery

The tank itself is also acquiring new capabilities. Surveillance radars and night/all weather vision equipment are improving effectiveness against camouflaged targets and enhancing capabilities to fight in all conditions.[74] Computerized fire control is also now standard equipment as are new anti-tank munitions both of which increase tank lethality. To counter the threats from tanks and other anti-tank systems, the Soviet Union has now fitted reactive armor which explodes when struck, thereby destabilizing projectiles and dissipating their energy. This has called into question the effectiveness of many types of NATO anti-tank weaponry but opinions are divided on the seriousness of the problem.[75] Various NATO nations are investigating fitting tanks with this type of armor and active armor which detonates just before impact. The United States has already begun to fit a new type of armor on its most recent Abrams M-1 tanks. This armor consists of a mesh made of depleted uranium encased in steel and is believed to be sufficiently resilient to prevent penetration by all types of Soviet anti-tank missiles including those known to be in development but not yet deployed.[76]

Turretless tanks are also being investigated although none is firmly scheduled for development. Such tanks would use automatic gun loaders, dispensing with one of four crewmembers and allowing the remainder to occupy only the hull, greatly reducing the turret volume to present a smaller target. Whether the reduction in crew and possible diminution of field of view are worthwhile is unclear, however. Each crewmember can spend up to 25 percent of his time on maintenance activities so automatic gun loaders may ultimately be used to increase operational efficiency without necessarily reducing crew size.

Also under investigation are composite armor to reduce weight and ceramic engines to increase fuel efficiency and also save weight. Research is also being conducted into liquid propellant and electromagnetic systems for tank armament and artillery. Liquid propellant research is in progress in France, the Federal Republic of Germany, the United Kingdom and the United States and is showing considerable potential. Liquid propellant is cheaper than solid propellant, less bulky, and gives a smoother acceleration resulting in less barrel wear. It is also insensitive to shock and not flammable at atmospheric pressure while being simpler to use in automated systems yielding a higher rate of fire.[77]

Electrically powered guns of various types are also being investigated. Depending on the outcome of tests scheduled for 1989, these could be in service in some form in the early 1990s. Largely as a result of research under the auspices of the Strategic Defense Initiative, substantial advances have been made in electrically powered guns to the extent that several types of tank-like guns are to be tested against each other in 1989, and against conventional guns in 1991. Many difficulties must be overcome, most notably the development of small, very substantial power supplies, but the potential benefits are considerable. Projectile velocities of up to six kilometers per second—any faster and air friction incinerates the projectile—yield greater range and permit the use of very small projectiles needing no explosive charge. It may even be possible to use such guns against both tanks and helicopters.[78]

Another novel weapon system which may find battlefield application further into the future is the weapons laser. Low-powered lasers are already used extensively for range finding, target designation, and training, but lethal systems would have to be several orders of magnitude more powerful. However, preliminary research in, for instance, the Federal Republic of Germany suggests that laser weapons may be useful against aerial targets after the turn of the century. If the many technical difficulties can be solved, such systems would offer a range of several kilometers, exceptional accuracy, and a very short engagement time.

Robot Vehicles

Well before the end of the century, semi-autonomous and autonomous vehicles will undoubtedly be employed on the battlefield. Several prototype systems are under development in the United States and preliminary tests have been encouraging. One program is the Teleoperated Mobile Antiarmor Platform (TMAP) scheduled for extensive testing in late 1988.[79] The TMAP will consist of a four-wheeled, 300 kilogram vehicle armed with four anti-tank missiles and a machine gun. The vehicle will be controlled by a single operator up to four kilometers away using a backpack-sized control unit and a radio or fiber optic link. Other similar concepts are under investigation to establish the feasibility of using robot vehicles for sentry duties or as more capable anti-tank vehicles. Such systems could be fielded in the early 1990s and might be followed by fully autonomous vehicles which could be used to hunt enemy vehicles in designated areas without human intervention. This type of capability will require several major technological advances, notably in image recognition, autonomous navigation, and autonomous terrain negotiation, but research in these areas is progressing well such that by the end of the century robot fighting vehicles are likely to be a viable proposition.

Conclusion

Technological progress is increasing the "smartness" of virtually all types of munition, and is facilitating the large-scale integration of complex systems. Provided such weapons and systems perform as planned, these will bring about substantial improvements in Western combat effectiveness. In addition, the trend towards smart weapons should help to achieve a better ratio between combat "teeth" and logistic "tails". The trend towards increasing smartness also opens up new opportunities for systems "trade-offs". In the past, for instance, traditional wisdom has held that the tank is the most effective anti-tank weapon, a conviction only briefly shaken by the advent of the portable and non-portable anti-tank missile. In the future this view is less likely to prevail as smart weapons ranging from tube and rocket artillery projectiles to mines and missiles such as FOG-M and TRIGAT plus airborne anti-armor weapons become more widespread. Consequently, it is more important than ever to avoid pressure to pursue a system-for-system replacement philosophy and instead to think in terms of the conduct of missions.

In doing so, however, the problems associated with complex system integration must be given due weight. If reliance is to be made on complex interactions between surveillance systems, communications and data processing, and delivery systems, reliability is vital as are interoperability, redundancy, and standardization. Unless attention is paid to these factors, the much-vaunted synergy these systems offer will give way at worst to impotence or at best to a fragmented, inefficient response to aggression.

Indeed, some analysts have raised substantial doubts about the performance of smart and complex integrated systems in the face of an ingenious and uncooperative foe, and have therefore cautioned against being seduced by these new technologies. Certainly it is true that "dumb" systems provide few if any opportunities for enemy deception while "smart" systems always provide some. In other words, little can go wrong with dumb systems whereas smart systems can suffer reliability problems and invite hostile efforts to counter them. Without going into the detailed technicalities of the arguments, however, the "naysayers" have raised many important issues. Among these are a history of inadequate and unrealistic testing of advanced systems and the prospects that an adversary will probably develop countermeasures which may have unpredictable consequences. On the first point, as noted earlier, test performance and actual performance can differ by a factor of ten such that there is a clear need to ensure that tests are conducted in sufficient numbers and in as realistic conditions as possible. Although this is expensive, to do otherwise with advanced technology weapons is a false economy. As regards countermeasures, there is no doubt that complex and automatic systems run the risk of enemy deception which cannot be wholly foreseen. This suggests that system redundancy and variation are essential and that it is prudent to retain systems which keep the "man in the loop". Fortunately, systems such as the fiber optic guided missile which keep the man in the loop but place him further out of harm's way are becoming available and indications are that this technique could be extended to systems with ranges up to 100 kilometers. Clearly, this promising technology merits particular emphasis.

In sum, new technologies offer a host of new opportunities for battlefield operations. But their introduction should be planned to allow interaction and substitution of systems, and reliability should be a paramount consideration. In other words, a prerequisite for coordinated, cohesive, and effective strategy on the battlefield is a coordinated, cohesive, and effective strategy for procurement.

4

AEROSPACE TECHNOLOGY

Aerospace technology—embracing fields such as aerodynamics, propulsion materials, and avionics—is exploiting a wide range of developments to produce systems with substantially improved performance. Lessons regarding such problems as poor reliability, troublesome maintenance, and cost escalation seem to have been learned, although it will be some time before these filter through to the operational level. This chapter examines the principal advances in aerospace technology, addressing the subject under three main headings; aircraft, missiles, and space systems. In each case, the focus is on projects in development and due to be fielded in the 1990s, but attention is also paid to some key "technology demonstration" projects which show particular promise.

Aircraft

The American F-15 originated from studies beginning in 1966 and the origins of the F-16 and F-18 date back to 1969. Since then, aerospace technologies have leapt forward so that these aircrafts' successors now in development in the United States and Europe will display dramatic improvements in both performance and reliability.

Some of the new aerospace technologies such as fly-by-wire and stealth are already incorporated in aircraft now flying, but none integrates all the technologies now available or, of course, those in advanced development.

Aerodynamics and Control Technology

In terms of aircraft handling, the most remarkable innovation is the successful development of "relaxed stability", already a feature of the F-16 and destined to be included in all future fighter aircraft. Traditionally, in aircraft design the center of gravity is further forward than the center of lift and the aircraft's tail produces a downward force to prevent the aircraft's nose from dipping. This configuration is inherently stable so that the aircraft has a natural tendency to fly straight and level. The disadvantage is that some of the wing's lift serves only to compensate for the downward force generated by the tail. At supersonic speeds, this penalty is even worse because the center of lift moves backwards, further away from the center of gravity, making the aircraft even more nose heavy and requiring more downward force to be exerted by the tail. The net effect is to waste engine power to maintain aircraft altitude. Unfortunately, the theoretical solutions to this problem—placing the center of gravity behind the wing's center of lift and using the tail to generate lift, or co-locating the centers of gravity and lift and using the tail for control only—are inherently unstable. If the aircraft moves even slightly nose up or nose down, the deviation becomes uncontrollable.[80] (A loose analogy is a "top heavy" ship which can remain upright in a flat calm but which would capsize if tilted even slightly).

To solve this problem, aircraft use "fly-by-wire" and "active control" technologies. The former replaces mechanical linkages and hydraulic actuators with electronic links and electromagnetic actuators which are lighter and more responsive. Active control technology links this with sensors to measure control commands, the state of control surfaces, airspeed, and angle of airflow, while computers tie the whole system together. In effect, the computers interpret the pilot's commands and then adjust the aircraft's control surfaces to achieve the desired effect. Stability problems are overcome by the computer which makes trimming adjustments to control surfaces about forty times a second. Without the computers performing this task, the aircraft would simply be uncontrollable, so multiple computers are fitted.

In the F-16, which has a traditional "tail aft" configuration, the tail generates lift in subsonic flight and at supersonic speeds—when the centers of gravity and lift are closer together—the tail essentially maintains balance.

Eventually, "fly-by-wire" will probably be replaced by "fly-by-light" as fiber optic cables offer higher data rate capacities than traditional wiring and are immune to phenomena such as electromagnetic pulse and electronic interference.

Active control technology (ACT) has several other desirable features beyond simply reducing drag. The aircraft's instability endows it with remarkable manoeuvrability, an obvious bonus. It also facilitates "carefree manoeuvring" whereby the computer will not permit the pilot to conduct manoeuvres which would destroy the airframe or render him unconscious through a phenomenon known as "g-induced loss of consciousness" (GLOC). This liberates the pilot from constantly monitoring his "g" indicator, allowing him to deal with other tasks in the sure knowledge that he can fully exploit his aircraft's capabilities.

ACT also permits smaller vertical stabilizers to be used, reducing drag still more. Another ACT feature is automatic diagnosis of control faults and battle damage plus the capability to compensate for them. Faults or damage which would render a traditional aircraft unflyable can be tolerated as the control computers automatically reconfigure control surfaces to compensate.

The F-16 was the first aircraft to incorporate ACT and will probably be the last fighter to employ the traditional "tail aft" layout. Designers in the United States, Europe and Israel all now favor canards forward of the wing which when coupled with ACT confer still more manoeuvrability. This manoeuvrability, however, is qualitatively different from that of traditional aircraft in that aircraft attitude is decoupled from its flight path. In other words, the aircraft will be able, for instance, to fly straight and level while the aircraft points to one side or up or down, to turn without banking, or to move sideways or vertically without turning or tilting.[81] This capability will facilitate weapons aiming and will be especially useful in aerial combat. Canards also can augment wing lift such that total lift is 30 to 50 percent greater than that produced by the canards and wing acting separately, and canards also can increase sustained turning rates—a measure of agility—by 10 percent.[82]

Further substantial improvements in aircraft performance will be realized by the "mission adaptive wing", a technology now being developed in the United States. The mission adaptive wing (MAW) has no leading and trailing edge flaps or ailerons but instead varies its profile. The wing has smooth, continuous surfaces and its curve can varied by small, built-in actuators. Tests on a modified F-111 have yielded range improvements of 25 to 30 percent (depending on flight profile), increased the aircraft's ceiling by 15 percent, and improved sustained turn rate by 20 percent. In initial tests, wing curvature was controlled manually but automatic adaptation is now being developed. An additional bonus is the capability to vary the profile along the wing so that at high loads more lift is generated closer to the wing root, reducing structural stress.[83]

Materials

Aircraft construction materials are also contributing to better performance. The use of composites—glass, boron, plastic or carbon fibers in an epoxy resin—is already widespread. For instance, 10 percent of the F-18 and 26 percent of the AV-8B by weight are constructed of composites.[84] Composites are lighter and stronger than metal alloys and are resistant to corrosion and cracking. Also, fractures do not propagate easily and resistance to bending can be concentrated in a particular direction by appropriate orientation of the fibers. This feature has made possible the X-29 technology demonstrator aircraft whose forward-swept wings are constructed from composites.

Existing composites are not free from drawbacks. Reactions between some composites and some metals can cause metal corrosion, and composite manufacture is a skilled, time-consuming, and expensive process. Composite structures must be cured in an autoclave at high pressures and temperatures, a slow process needed to produce the irreversible chemical reactions which harden the epoxy resin. Fabrication faults cannot be corrected so parts must be scrapped. Consequently, considerable attention is devoted to quality control.

New composite materials, however, will solve these problems. These materials are thermoplastics which are fluid at high temperatures. These can be retreated and reformed to correct faults or repair damage, and they lend themselves well to traditional mass production techniques such as injection moulding and vacuum forming. They can also be welded using existing ultrasonic techniques and bonded effectively with a wide range of powerful adhesives. Consequently, the costs of these materials should be up to 50 percent lower than those requiring autoclaving. Furthermore, they are even tougher and more impact resistant, and those incorporating carbon fibers provide shielding against electromagnetic interference and are particularly effective in absorbing radar signals.[85]

Metallic alloys are also being improved. Aluminium-lithium alloys have shown promise for decades but only recently have ductility deficiencies been solved, probably by introducing trace quantities of metals such as zirconium. New aluminium-lithium alloys are now being used which are five percent stronger and nine percent lighter than traditional aluminium alloys. A further development is fiber reinforced alloy which offers a 25 percent weight reduction.[86]

Propulsion

Propulsion technology too has advanced so that engines on next-generation aircraft will be lighter, more powerful, more efficient, more durable, and easier to maintain. New materials and production techniques will allow engines to run hotter and therefore more efficiently and also to incorporate fewer parts. Turbine blades, for instance, can be produced from a single crystal which adds greatly to their durability.

Three-dimensional computer design techniques are also leading to new turbine blade shapes which are shorter, and more twisted—rather like marine propellers—rendering them more efficient and less susceptible to damage. Vectored thrust is also to be incorporated in new American fighter engines, permitting deviations of 20 to 30 degrees in thrust elevation, a technique which improves manoeuvrability and short take-off and landing (STOL) capability. Dornier and Northrop are also investigating the use of vectored thrust to replace the traditional horizontal tail on advanced fighters. This would increase STOL performance, reduce drag, and decrease radar signature.

These developments add up to engines with thrust-to-weight ratios of 10:1 without afterburners, compared with 7.5:1 on current engines so that aircraft will have supersonic persistence, the ability to cruise efficiently at supersonic speeds. This extra thrust also improves manoeuvrability because in tight turns aircraft drag increases dramatically, slowing down the aircraft accordingly; extra thrust allows speed to be maintained during manoeuvres, a feature more valuable to a fighter than a simple increase in maximum speed.

Examples of such engines are the Pratt and Whitney F-119 and the General Electric F-120 now under competitive development for the United States Advanced Tactical Fighter, and the EJ-200 being developed by a European consortium for the European Fighter Aircraft. The goals for all these engines are improved thrust-to-weight ratios, efficiency, and simplicity. The engines for the Advanced Tactical Fighter include a 300 percent increase in reliability and 50 percent decreases in weight and parts compared with the engines on the F-15. Goals for the EJ-200 are similar and it is already established that whereas the RB-199 which powers the Tornado employs three spools and 16 stages, the EJ-200 will employ only two spools and 10 stages with each stage having fewer aerofoils.

To maximize engine performance, control will be electronic rather than manual. "Full authority digital engine control" uses engine and aircraft sensors coupled with a computer to operate the engine as efficiently and as close to the edge of the engine's performance envelope as possible.[87]

Avionics

Avionics are another critical factor in improving aircraft capability. Indeed, avionics are the largest budget item on new fighter aircraft. As well as the sophisticated electronics needed for active control technology, new fighters will incorporate avionics of unprecedented capability both to enhance capability and reduce pilot workload. The F-14A cockpit contains over 400 switches and dials and is typical of current advanced aircraft.[88] New electronics technology now permits rationalization of cockpit layout so that television-type displays using graphic techniques can reduce pilot workload by presenting data in easier to read formats when requested by the pilot or when in-built electronic systems "judge" it necessary. Conventional head-up displays can also be supplemented by helmet-mounted systems such as the American "Agile Eye" so that, for instance, a virtual image of terrain features or infrared images can be superimposed on the pilot's field of view, and weapon aiming can take place from cues provided by the helmet's orientation, dispensing with the need to "aim" the aircraft.

Pilot workload will be further reduced by the use of voice command and audible cues. In France, Crouzet has made great progress with voice control to liberate the pilots' hands from tasks such as flicking arming switches and several companies are developing speech synthesizers which will be able to take the place of—for instance—SAM attack warning lights.[89]

Aircraft radar is also being improved by moving to phased array systems which scan electronically rather than mechanically. These can track individual targets while simultaneously searching for new ones and are also more fault tolerant. Increasing use is also to be made of infrared systems such as the American Low Altitude Navigation and Targeting InfraRed system for Night (LANTIRN) which incorporates an infrared imaging system with a wide field of view, terrain-following radar, and a laser designator. LANTIRN can produce images on a television display or a head-up display.[90]

Aircraft navigation is another area which is already seeing dramatic developments. Systems such as the British Terrain Profile Matching (TERPROM) system contain digital maps of several hundred thousand square kilometers of terrain in a unit weighing only 15 kilograms. This uses a low-emission radar altimeter to measure terrain profile which is then compared electronically with the map to locate the aircraft very precisely and, if required, to fly the aircraft along a desired route while hugging ground contours.[91] The use of laser altimeters will reduce emissions still further so that aircraft will be able to navigate with extreme precision without betraying their presence with electronic emissions.

Considerable attention is being paid to improving other avionic systems associated with electronic countermeasures (ECM), electronic warfare, communications and identification friend or foe (IFF), and to integrate avionics systems as far as possible so that antenna and processing equipment can be shared. Examples include the American Integrated Electronic Warfare System (INEWS) and the Integrated Communications Navigation and Identification Avionics (ICNIA) programs.[92]

All these systems are to be designed with reliability and ease of maintenance in mind and extensive use will be made of VHSIC technology to reduce systems volume, weight, and power requirements. ICNIA, for instance, should achieve a mean time between failures of 1,233 hours compared with 20 hours for discrete equipment. It would also reduce the number of distinct module types to only 28 compared with the 1,300 that would be required using discrete communications, navigation, and identification equipment.[93]

Software production techniques are also being refined to handle the enormous challenges posed by such sophisticated electronics and already great progress has been made. In the United Kingdom, for instance, improvements in software production have reduced errors from 15 per thousand lines of code on the Tornado air-defense variant to two per thousand on the Experimental Aircraft Program, while production per man-year has increased by a factor of 34, and man-hours on testing have declined by a factor of seven.[94]

In the United States, the PAVE Pillar Program is investigating advanced avionics architecture (A3) for application on several types of planned aircraft including the Advanced Tactical Fighter. This is to exploit technologies such as VHSIC, multi-purpose colour displays (MPCD), fiber optics, speech recognition and speech synthesis, in order to produce a total aircraft electronics system which will be both highly capable and less demanding to operate. The intention is to allow the pilot to maintain "hands on throttle and stick" (HOTAS) conducting tasks such as selecting and aiming weapons, choosing radio frequencies, and requesting aircraft and sensor status orally, and having data presented as required on display screens and by audible means. At the same time, aircraft avionics will be made more reliable and easier to maintain by incorporating built-in test equipment, fault-tolerant electronics, and easily replaced electronics modules. Compared with an F-16's avionics unit which has 86,600 connections and 58 line replaceable units, PAVE Pillar will produce a suite with 56,000 connections and 28 line replaceable units. Mean time between failures will be increased from 7.3 hours to 35 hours and the proportion of maintenance time devoted to avionics will decline from 17 percent to five percent. Consequently, the aircraft sortie rate will rise from 3.1 per aircraft day to 4.5 per aircraft day.[95]

Stealth

One of the most radical innovations for new aircraft is stealth technology. Stealth is not a simple device but in fact is a combination of design, performance, and materials. Large, flat surfaces which reflect radar well must be avoided in favor of smooth, blended contours which disperse a radar signal. This includes, where possible, internal weapons stowage. Engine inlets and turbine blades must be recessed as far as possible—some aircraft radars such as the Hughes

APG-70 can process radar returns from targets almost 150 kilometers distant to identify the target by the number of turbine blades—and engine exhausts must be recessed and mixed with cooler air to reduce infrared emissions.[96]

As regards materials, composites are very poor radar reflectors and are used as extensively as possible, notably on vertical stabilizers. Radar absorbing materials (RAM) have also been developed to coat surfaces which otherwise would be radar reflective. One type of RAM is believed to consist of polymers impregnated with minute metallic spheres which absorb a great deal of radar signal energy and ensure that any residual reflections are dispersed.[97] Another type which is 90 percent lighter but nevertheless cuts radar reflections by 80 percent, uses certain salts in a polymer base.[98] Other RAM no doubt exist but few details are publicly available.

Another aspect of stealth technology is decreased electronic emissions through, for instance, passive navigation, spread spectrum communications and—as far as possible—passive instead of active sensors.[99] Also playing their part are intelligence, tactics, and speed. Knowledge of radar sites and terrain masking features facilitates the planning of routes to minimize detection, and the faster an aircraft travels, the less time it dwells in a radar's area.

The application of all these techniques can dramatically reduce an aircraft's visibility to radar. The American B1B, for instance, has only 10 percent the radar cross section of the earlier B1A (which did not enter production) and only one percent the radar cross section of some types of B-52.[100] The American stealth fighter reportedly has an exceptionally low radar cross section though, as the first generation of truly stealthy aircraft, some sacrifices have probably been made in agility, armament, and performance.

Stealth technology does not render an aircraft invisible to radar but substantially decreases the range at which a particular radar is likely to detect it. Very roughly, a tenfold reduction in radar cross section cuts detection distance by half.[101] Surveillance radars which usually operate at longer wavelengths are the least susceptible to deception by stealth, but shorter wavelength systems such as those associated with SAM sites and those on SAMs themselves are much more susceptible.

New Fixed-Wing Aircraft

These types of technologies will be embodied in new aircraft now in development on both sides of the Atlantic, scheduled to enter service in the mid-1990s. In the United States, the Advanced Tactical Fighter (ATF) is being developed as a replacement for the F-15 and possibly for the F-14D. The Advanced Technology Aircraft (ATA) is also being developed as a possible replacement for the A-6 and the F-111. The ATF will be extremely agile, reaping the benefits of active control technology, the mission adaptive wing, and advanced propulsion technology including vectored thrust. It will also have a short take-off and landing capability and, through its sophisticated electronics and armament, is expected to enjoy "first detect, first kill, multiple engagement" capability. It will cruise at Mach 1.5 without afterburner and, compared with the F-15, will have twice the combat radius, twice the mean time between failures, and will require half the maintenance personnel. It should also have a radar cross section between one percent and 10 percent of the F-15's.[102]

Other American fixed-wing projects include two or perhaps three stealth aircraft. The B-2 stealth bomber is known to be under development, a smaller multi-purpose stealth aircraft is believed to be in production and a long-range stealth reconnaissance aircraft project may be in hand. Details of all stealth projects are highly classified, so their status—particularly regarding the latter two—is difficult to establish.

The European Fighter Aircraft (EFA) being developed by the Federal Republic of Germany, Italy, Spain, and the United Kingdom will be rather less sophisticated than the ATF but certainly more so than the F-16. It will enjoy all the advantages of active control technology, advanced propulsion, and sophisticated electronics. It will be exceptionally agile and, although less stealthy than the ATF, will nevertheless be designed to minimize radar cross section. As with the ATF, reliability and ease of maintenance are ranked equally with performance as design goals.[103] Another European project is the French Rafale, broadly similar to EFA but somewhat lighter.

New Rotary-Wing Aircraft

Rotary-wing aircraft are benefitting from essentially the same types of new technologies being applied to fixed-wing aircraft. Fly-by-wire and fly-by-light are reducing the weight of control systems, and active control technology is being investigated as a means of enhancing agility although in a more limited sense. Relaxed stability has no helicopter analogue, but active control technology can nevertheless give a pilot "carefree manoeuvring". Studies indicate that only 50 percent of a helicopter's agility is actually exploited because flying such machines is so demanding and it is extremely difficult to take full advantage of the helicopter's capability. Active control technology will reduce the pilot's workload and ensure that the helicopter flies as efficiently and as close to the performance envelope as possible, thereby increasing achieved agility up to 90 percent of available agility.[104]

Composites are an especially useful development for helicopters where weight is at a premium. Their use is increasing to the extent that about 80 percent of the American V-22 Osprey airframe, for instance, will be constructed of composites. It is unlikely that any new helicopters will employ metal rotors since composites are lighter, more damage resistant, and require virtually no maintenance. Their directional resistance also facilitates novel rotor designs such as the British Experimental Rotor Program (BERP) which provides 30 percent more thrust and adds 3,000 meters to the ceiling while increasing cruising speed to above 370 kilometers per hour.[105] Rotor hubs are also incorporating new materials which allow the number of parts be reduced dramatically, yielding other advantages such as reduced drag, increased reliability and easier maintenance.

Propulsion systems are becoming more powerful, lighter, simpler, and easier to maintain. Reliability—an especially important feature in helicopter propulsion systems—is also increasing, as is the use of automatic fault diagnosis.

Helicopter avionics are also advancing. Infrared imaging systems are permitting operations at night and in poor weather, and audible warning systems using synthesized speech are beginning to appear.[106]

These types of technologies are being incorporated in such systems as the Italian "Mangusta" due to enter service in 1988. Design features include minimum silhouette, low rotor flicker, reduced reflection, extremely low infrared signature, invulnerability to 12.7 millimeter projectiles, self-sealing fuel tanks, excellent crash-worthiness, and a sophisticated avionics suite incorporating multi-function displays and automatic fault diagnosis. Air-to-air missiles can be carried in addition to anti-tank missiles, 20 millimeter cannon and unguided "salvo" rockets, reflecting growing sentiment that future combat will require the capability to deal with threats from rotary-wing and fixed-wing aircraft.[107]

Another new system is the Anglo-Italian EH-101 due to enter service in an anti-submarine warfare role in the early 1990s. The EH-101 will have an eight hour endurance, and double the range and payload of its predecessor the "Sea King". It will be equipped with sonobuoys, dunking sonar, magnetic anomaly detectors, a pulse-compression radar designed to spot submarine periscopes and snorkels, and the GEC AQS-903 acoustic processing and display system which is eight times as powerful as the AQS-901 fitted to the British "Nimrod" maritime patrol aircraft.[108]

Other European helicopters in various stages of planning and development include the Franco-German PAH-2 (an advanced attack helicopter intended to match the performance and lethality of the American AH-64 but in a smaller package), the "Tonal" (a prospective successor to the "Mangusta" to be developed by Italy, the Netherlands, Spain and the United Kingdom), and the French multi-purpose Panther.[109]

The largest helicopter program planned within NATO is the American Light Helicopter Experimental (LHX) due to enter service in the mid-1990s. Until defense budget cuts forced a revision of the program in early 1988, the LHX was to have been produced in two versions, scout and attack (SCAT) and utility, with about 70 percent parts commonality. The revised program dispensed with the utility version but retained the scout and attack version. The target weight is about 3,500 kilograms and other specifications include a conventional design, a speed of 315 kilometers per hour, ferry range of 2,300 kilometers, high manoeuvrability, and low observables. The avionics suite will be very sophisticated being, as far as possible, common with that of the Advanced Tactical Fighter. Features include computer-aided target recognition, video instruments, helmet-mounted display, interactive voice control, night vision equipment, and automatic fault diagnosis. Armament will include four Hellfire air-to-ground missiles, two Stinger air-to-air missiles, and a 20 millimeter cannon. Also included in the specification is a 30 to 40 percent reduction in operations and maintenance costs over current models.[110]

Original plans called for 5,000 LHXs to replace the United States Army's 7,000 AH-1s, UH-1s, OH-58s, and OH-6s now in service, some of which date back to the 1960s. Changing specifications—such as the vehicle's weight and armament—reduced this number to just over 4,000 at a total cost of about $66 billion and then pressures on the defense budget led to a further reduction such that current plans call only for 2,096 SCAT variants at a total cost of $37 billion.[111] An issue which has caused substantial difficulty is whether the LHX should have one or two pilots. To resolve this, prototypes will be produced for both configurations. Also controversial is the choice of a conventional design over a tilt-rotor arrangement which could meet or exceed all specifications according to some sources.[112] Provided LHX's weight specification does not rise—and it has risen and fallen in the past—the conventional design seems likely to be retained.

The Tilt-Rotor Aircraft

The tilt-rotor helicopter concept, nevertheless, certainly has a future elsewhere in the American armed forces.

The tilt-rotor concept offers the flexibility of the helicopter and a much higher maximum speed. Conventional helicopters have a maximum speed of about 370 kilometers per hour because beyond that the retreating blades stall. In other words, the blades on one side of the helicopter travel "forwards" and those on the other side travel "backwards" so that their pitch has to be adjusted to maintain constant lift on either side of the helicopter. At high speeds, however, no amount of pitch adjustment can compensate for this asymmetry. The tilt-rotor vehicle overcomes this limitation by having two rotors, one at each end of its wings. For take-off and landing, these are positioned to provide vertical lift but once airborne they can pivot through 90 degrees to act as aircraft propellers, albeit very large ones.

The V-22 Osprey due to enter service in 1991 is such a vehicle. Constructed largely from composite materials and employing fly-by-wire controls, the V-22 with its rotors generating vertical lift will travel at about 185 kilometers per hour forwards and about 60 kilometers per hour sideways or backwards. With its rotors tilted fully forwards it will be able to reach speeds of over 500 kilometers per hour.[113] With an empty weight of around 14,500 kilograms, the V-22 is clearly much larger than the LHX. It will carry 24 fully equipped troops or a 4,500 kilogram payload anywhere within an operational radius of around 1,000 kilometers and work is already in hand to increase range to meet requirements for special operations.[114]

The missions planned for the V-22 include combat support, medical evacuation, search and rescue, and transport for special operations. In these missions it would replace the Marine Corps' CH-46E Sea Knight and CH-53A/D Sea Stallions, the Army's CH-47 Chinooks, the Navy's SH-3D Sea Kings, and the Air Force's HH-53 and C-130 used in special operations. Also proposed are gunship versions and a naval variant, the SV-22, to be used for anti-submarine warfare. This latter type would be fitted with a dipping sonar, 60 sonobuoys, a multimode radar and ASW torpedoes.[115]

Cuts in the defense budget have caused some revisions to the acquisition program. Current plans call for 682 production aircraft with 552 destined for the Marine Corps, 80 for the United States Air Force (equipped with more sophisticated avionics and extra fuel capacity for special operations), and 50 for the United States Navy (to be used for combat, search and rescue, special warfare, and logistics support). The proposal to acquire 350 ASW versions to replace S-3 Viking aircraft has been dropped as has the Army requirement for 231 V-22s in the basic Marine Corps configuration. These may be reinstated once the V-22 has proved itself in other roles and when room can be found within the relevant service budgets.[116] The fly-away cost of the V-22 is about $16.6 million (in Fiscal Year 1986 dollars) but if test equipment, spares, training, simulation and non-recurring research and development funding is included, the total cost would be about $30 million per copy.[117]

The V-22 will offer many operational advantages. It will be more fuel-efficient than the helicopters it replaces and will be easier and cheaper to maintain. It will also be four times quieter, more resistant to battle damage, and less vulnerable to heat-seeking missiles due to exhaust suppressors and wing-tip mounted engines which do not bathe the fuselage in hot air, features which yield an infrared signature one sixth that of a comparable helicopter. Crash survivability is also enhanced by the wing-tip mounted engines which ensure that hot metal and rotors do not fall onto the occupants after impact and a low center of gravity which will reduce the likelihood of the V-22 rolling over whether or not its wings remain attached to the fuselage.[118]

For the United States Marine Corps, the V-22 offers double the speed and five times the range of the CH-46 so that a parent vessel will be able to stay further out to sea and still launch an amphibious assault more rapidly and further afield than is currently possible. This enhances the survivability of the parent vessel by placing it further from hostile forces ashore and also increases the effectiveness of the forces it deploys since they can build up more quickly and their landing options are more extensive and more difficult for an opponent to predict.

To cite figures for just one operational benefit, a reinforced battalion can be lifted ashore from a range of 280 kilometers in about ten hours using CH-53 and CH-46 helicopters: using the V-22, the same feat could be achieved in two hours.[119] In view of performance enhancements of this nature, the V-22 seems set eventually to supersede traditional heavy helicopters.

Technology Demonstration Projects

In the aviation area, many other research projects are in hand to investigate promising technologies for the future without being tied specifically to the development of a specific project. In the United States, for instance, work is progressing on a rotary-wing concept which is radically different from conventional and tilt-rotor vehicles.

Tilt-rotors are only one way to overcome the speed limitations of rotary wing aircraft. Another solution is the "advancing blade" concept which uses contra-rotating rotors. This technique has been investigated in the West but the only operational helicopter of this type is the Soviet "Hokum". The high drag and vibration problems limit speed to about 460 kilometers per hour, rather more than a conventional helicopter but less than a tilt-rotor.

Another promising possibility now being investigated in the United States is the "X-wing" vehicle which should attain speeds of about 1,000 kilometers per hour. The X-wing concept is being tested in a $200 million—originally $100 million—program and the relevant technologies embodied in a demonstration vehicle should be established by the early 1990s.[120]

The X-wing's rotor is mounted conventionally but its blades are extremely broad and have a symmetrical cross-section. Air is blown through slots which run the length of the blades on both leading and trailing edges, generating lift through a phenomenon known as the Coanda effect. Once sufficient forward speed has been attained, the rotors are stopped such that each acts as a wing angled at 45 degrees to the fuselage. The rotors angled forward must cope with considerable stress so the rotors are extremely rigid, composite structures.

The technique sounds relatively simple but requires very sophisticated control of air circulation (pneumodynamics) in the rotor system, particularly during transitions between rotary-wing flight and fixed-wing flight. Indeed, the research vehicle now being developed embodies about as much computing power as the American space shuttle.[121] Another requirement for an operational X-wing is an engine which can deliver shaft power for helicopter flight and thrust for high-speed forward flight. This can be done but the current research vehicle uses both types of engine to reduce program risk.[122]

Once the key technologies have been demonstrated and provided costs prove acceptable, the X-wing could have enormous potential, combining the versatility of the helicopter with the performance of a high-subsonic aircraft.[123]

The United States is also testing a novel jet aircraft with forward swept wings, the X-29A. Forward-swept wings offer the delayed transonic drag of rearward-swept wings plus reduced stalling speed, improved low-speed handling, and the prospect of increased speed and agility for any given installed power.[124] Only the advent of composite materials allowed the concept to be verified since these can be used to construct wings strong enough to resist the extremely strong aerodynamic forces acting on a forward-swept wing. Two X-29A research aircraft are flying and tests have produced very encouraging test results so some future fighters might well employ the forward-swept configuration.

In addition, the United States and the Federal Republic of Germany are jointly building two research aircraft in the Enhanced Fighter Manoeuvrability project also known as the X-31A. Due to fly in mid-1989, these aircraft will have vectored thrust and advanced control features to enable safe manoeuvring at high angles of attack and speeds which would cause traditional aircraft to stall. The results derived may be applied to the European Fighter Aircraft, the Advanced Tactical Fighter, the Advanced Tactical Aircraft, or to projects further into the future.[125]

The United Kingdom and the United States are investigating propulsion technology for supersonic vertical take-off and landing aircraft. The AV-8B/Harrier II has been a major success and the next logical step is to proceed towards a supersonic successor.

The Supercockpit

Other areas of research include conductive composite materials which will allow antenna to be built into an aircraft's skin, high energy aviation fuels, and advanced cockpit designs. With respect to the cockpit, arriving just too late for immediate incorporation in the ATF will be the "super cockpit" or "virtual cockpit", expected to be available by 1996. The super cockpit will consist, in part, of a helmet which will completely cover the pilot's head and which will generate a three dimensional, graphic representation of the view from the cockpit which can replace or supplement the actual view. Helmet position will be monitored so that the image projected will shift according to head position giving the pilot the impression that he is "inside" the computer-generated image. Terrain features will be represented with complete clarity irrespective of external conditions. Objects of interest such as hostile radars, SAM sites, and friendly and hostile aircraft will be depicted along with colored cones representing the reach of their weapons, with a colored "pathway" showing the safest route to fly to the chosen target.

Engagements will be conducted by the pilot simply looking at a target—his object of interest being determined by sensors monitoring eye positions—and verbally selecting the weapons of his choice. Sensors and tactile stimulators in his gloves will enable the pilot to "feel" and operate computer-generated images of any controls and switches desired. Weapons and aircraft status information will also be called up verbally and superimposed on his view or delivered audibly as appropriate. In addition, the pilot will be able to change his field of view to "God's Eye" which will represent the scene around his aircraft from a point about 6,000 meters above it.

The other part of the super cockpit will be the "pilot's associate", also known as the "silicon co-pilot". This will be an artificial intelligence system equipped with a knowledge of tactics, weapons characteristics, and topography and capable of "fusing" information from the aircraft's own sensors with that provided by pre-flight intelligence and external sources such as AWACS and JSTARS. If needed, the pilot's associate will be able automatically to handle defensive systems such as electronic countermeasures, chaff and decoys, and to prepare weapons for firing. Furthermore, if sensors on the pilot detect hypoxia or g-induced loss of consciousness, the pilot's associate will be able to decrease altitude or "pull up" as appropriate.

Futuristic as it may seem, most of the individual techniques needed for such systems have been demonstrated already, including helmets to produce three-dimensional images, eyeball and hand sensors to facilitate "interaction" with these images, and speech recognition. Programs such as the Visually Coupled Airborne Systems Simulator (VCASS) are pushing the technology ahead and program "milestones" have been set. In 1989, head-aimed fire control is to be established; in 1991, the all-aspect head-up display; in 1994, the complete virtual cockpit; and in 1996, the virtual cockpit linked to an artificial intelligence pilot's associate. Initial indications derived from simulations are that pilots like this concept and will be able to use it extremely effectively.[126]

Missiles

Missile technology is advancing rapidly, primarily as a result of progress in electronics but also through progress in areas such as propulsion, aerodynamics, and stealth. As in other areas, emphasis is also placed on using new technology to increase reliability, decrease maintenance requirements, and reduce costs. In addition, modular designs are increasingly favored so that technical advances can be absorbed as they become available, a feature known in the United States as pre-planned product improvement (P^3I).

Advances in electronics are yielding rapid increases in capability and reliability accompanied by decreases in size. These features create obvious opportunities for missile design, allowing overall reduction in missile size, or increases in space devoted to some combination of propulsion, fuel, and warhead, or even opening up completely new types of performance. For example, a missile seeker may incorporate up to 40,000 individual sensing elements (although 10,000 is more common) but the advent of seekers with 100,000 elements is at hand and 1,000,000 is a realistic goal. As the number of elements increases along with data processing capabilities, the seekers' field of view, resolution, and sensitivity also increase so that targets can be identified at greater ranges or in adverse conditions and the missile can pick out the target's most vulnerable point for impact. Radar guidance systems too are advancing. The radar produced in 1968 for the F-14 weighed 141 kilograms, that of the F-15 in 1972, 78 kilograms and that of the F-18 in 1976, 50 kilograms; the radar on the new Advanced Medium-Range Air-to-Air Missile will have a similar capability as these aircraft radars but will weigh only nine kilograms. In addition, a trend towards shorter wavelength radars gives a narrower beam width which provides both better discrimination and improved resistance to electronic countermeasures.[127]

Propulsion technology has advanced on several fronts, greatly enhancing the performance of a wide variety of missiles. The small, relatively cheap and very efficient turbofan along with high energy fuels has facilitated the production of small yet very long-range missiles such as the Cruise and its successor the Advanced Cruise Missile, believed capable of achieving intercontinental range. Hybrid rocket/ramjet engines such as those on the French Air-Sol Moyenne Portée nuclear stand-off missile provide an excellent combination of size, weight range and speed, and a trend towards small air-breathing engines where rocket propulsion was once favored is roughly doubling missile range for a given propellant weight.[128]

With respect to air-to-air missiles, the main NATO programs are the Advanced Medium-Range Air-to-Air Missile (AMRAAM) being developed in the United States and the Advanced Short-range Air-to-Air Missile (ASRAAM) being developed by the United Kingdom and the Federal Republic of Germany. These complementary weapons will be used extensively on NATO aircraft from the early 1990s.

AMRAAM, also known as the AIM 120A, is a beyond-visual-range (BVR) weapon with a range possibly as high as 75 kilometers. Its predecessor, the AIM 7 Sparrow, uses semi-active homing which requires the launch aircraft to illuminate the target with its radar right up to impact. This means that the launch aircraft must continue to approach the hostile aircraft throughout the engagement. AMRAAM, however, has an active radar seeker which can lock on to a target independently at a distance which varies with target radar cross section, background clutter, and electronic countermeasures. The launch aircraft's radar only needs to provide periodic up dates to AMRAAM until the missile's own seeker locks on and so can handle several AMRAAM engagements simultaneously. Since AMRAAM's speed—in excess of Mach 4—is greater than that of its predecessors', the launch aircraft can break off sooner, probably before the targets can counterattack. AMRAAM can also be used as a "fire-and-forget" missile if the engagement takes place inside the range of AMRAAM's seeker of if the fire control system can predict a launch trajectory which will carry AMRAAM to within autonomous lock-on range. Plans are already in hand to replace guidance components with more advanced versions now available, and a "compressed carriage" version with folding wings and fins for internal stowage on the ATF is being studied. ASRAAM is intended for closer "dogfight" combat at ranges up to about 15 kilometers. This will be infrared guided and somewhat shorter and fatter than the AIM-9 Sidewinder it will ultimately replace.[129]

France has opted for a domestically produced missile, the MICA, which will use a common airframe but different guidance systems for the medium-range and short-range missions. For medium range, a guidance system similar to AMRAAM's will be used and for short range, a system similar to ASRAAM's. Maximum range will be between 50 and 60 kilometers and the missile will be somewhat smaller and lighter than AMRAAM.[130]

Another example which well illustrates advances in capability is the United States Navy study of an Advanced Air-to-Air Missile to replace the long range AIM 54 Phoenix. The goals include a reduction in weight—from 430 kilograms down to 135 kilograms—and an increase in range to more than 185 kilometers compared with the Phoenix's 150 kilometers. Guidance will be semi-active with the launch aircraft's radar using coded pulses for each missile so that several engagements can be managed simultaneously. Confidence in the planned guidance system is such that instead of carrying a large, proximity-fused warhead, AAAM is expected to hit its target directly and therefore will carry only a small warhead.[131]

For defense against surface-to-air missiles, aircraft carry anti-radiation missiles such as the American AGM 45 Shrike and AGM 88 HARM (High-speed Anti-Radiation Missile) which home in on hostile radars. A British system, the Air Launched Anti-Radiation Missile (ALARM) now under development, will ascend to a high altitude and descend slowly by parachute, searching for and analyzing radar signals. When it selects a target, it will release its parachute and home-in on the signal or, if the radar is switched off, will use an inertial guidance system to attack the transmitter's last location. The United States has begun to deploy a completely new anti-radiation missile, the "Tacit Rainbow" which resembles a small Cruise missile and which can loiter for several hours, waiting for suitable radars to be activated. Of course, such systems can be successful if they only succeed in causing hostile radars to shut down, allowing friendly aircraft to penetrate enemy airspace.[132]

A host of other air-to-surface weapons are available and in development. These range from simple unguided bombs to sophisticated homing missiles. The trends are towards stand-off weapons which are launched at some distance from the target so that the launch aircraft can avoid terminal defenses, and towards guided weapons which have a high probability of finding their targets. A widely-used guidance technique is to illuminate a target with a laser and launch a missile or gliding bomb which homes in on the reflected laser light. Other approaches include television or infrared techniques. A particularly interesting variation under investigation in the United States is the hyper-velocity missile for use against armored vehicles, filling the "gap" between aircraft cannon and the 20 kilometer range of the laser guided "Maverick" missile.[133]

These missiles would travel at about 1,400 meters per second (over 5,000 kilometers per hour) over a range of about five kilometers. The kinetic energy of the missile would be enormous—roughly equivalent to a car travelling at 750 kilometers per hour—so no explosive warhead would be necessary. Guidance would be provided by a laser designator which would support multiple engagements and target selection would be made using a wide field of view, forward looking infrared system. The missiles would be mounted in 12-round pods small and light enough for an F-16 or A-10 aircraft to carry two. The extremely rapid engagement time would permit the launch aircraft to break off about two kilometers distant from the targets. A particularly attractive feature should be the low price, approximately one tenth the cost of the Maverick.

Another especially noteworthy project is the Modular Stand-Off Weapon (M-SOW) now being developed jointly by the Federal Republic of Germany, Italy, Spain, the United Kingdom and the United States. This has arisen from an agreement to synthesize a variety of stand-off weapons projects into one "family" of weapons for use against tactical targets at various stand-off ranges. Current plans call for a sub-munitions dispensing version with a range of 30 to 50 kilometers for use against airfields and fixed air defense sites; a 185 to 600 kilometer version to carry either a unitary warhead or sub-munitions for use against rear area airfields, depots, and bridges; and a 15 to 30 kilometer version carrying "smart" sub-munitions for use against armored mobile targets. Although key features such as propulsion type and guidance technique have still to be established, several attractive options are available. Short and long range types may use different propulsion methods such as rockets for short range and turbofans for long range, and guidance could be inertial, supplemented by perhaps terrain profile matching or terrain contour matching both of which essentially compare terrain features with electronic "maps" and allow accuracies of a few tens of meters.[134]

Surface-to-air missiles (SAMs) too are reaping the benefits of technological progress. Again, a multiplicity of specific systems exists and still more are on the drawing board. SAMs vary from small, shoulder-launched types with a range of a few kilometers up to large, vehicle-mounted systems with a range of up to 150 kilometers. An example of the former is the British Starstreak due to enter service in the early 1990s. Starstreak will be a "hit-to-kill" system which will have three "darts" per round to increase kill probability and will have a speed of Mach 6.[135] An example of the latter is the American Patriot which entered service in 1984.[136] This has a maximum range of about 145 kilometers, a ceiling of 24,000 meters, a speed of up to Mach 6 and 40g manoeuvrability. Compared with older systems such as the Hawk, it is also easier to maintain, cheaper to operate, and requires fewer personnel. While the Hawk requires five radars, and two main vehicles to support six, three-round launches, the Patriot requires only one radar and one vehicle to support eight, four-round launches. The Patriot also requires only 765 personnel per battalion compared with the Hawk's 878, and 1,928 repair parts compared with 4,398.

Another feature of the Patriot is that with appropriate modifications—to guidance software for example—it is capable of intercepting warheads from short-range ballistic missiles.

In sum, the key trends in missile technology are towards reduced acquisition and operating costs through automated design and manufacturing, increased reliability, built-in test equipment and easier maintenance; increased stand-off range through advances in guidance and propulsion techniques; and a general heightening of overall capability.

Space Systems

Space systems play increasingly important roles in terrestrial military operations. Military satellites currently perform six main tasks—nuclear early warning and attack assessment, surveillance and reconnaissance, communications, navigation, meteorology, and geodesy—but in the West only the United States has systems to perform all these tasks. Other NATO nations employ satellites for military communications but plans are in hand in some countries to develop various types of surveillance and reconnaissance systems, and some American systems are (or will be) used by other NATO countries.

Nuclear Early Warning and Geodesy

Nuclear early-warning and attack assessment satellites and geodetic satellites can be addressed very briefly since they are of little relevance to conventional warfare. For nuclear early warning, the United States currently uses Defense Support Program (DSP) satellites in geostationary orbits over the Indian Ocean, the Pacific Ocean, and South America to detect missile launches from the Soviet Union, China, and the Atlantic and Pacific Oceans. These satellites use infrared sensors to locate a missile launch with an accuracy of about six kilometers. Fears about laser weapons which could blind or—if sufficiently powerful—disable such satellites have led the United States to develop a new generation of early-warning satellites for deployment in the early 1990s. Known as the Satellite Early Warning System (SEWS), these will "stare" constantly at their target areas instead of scanning them periodically and the sensors will be laser resistant. The satellites themselves will also be "hardened" to withstand powerful laser illumination.

Early warning satellites are also equipped with devices to detect nuclear explosions. In peacetime these sensors help verify compliance with the treaty banning atmospheric nuclear tests, while in a nuclear war they would facilitate rapid damage and attack assessment. From the late 1980s, however, these functions will be performed by the Integrated Operational Nudet (nuclear detonation) Detection System (IONDS), which will consist of sensors fitted to Navstar navigation satellites. This system will provide much more precise data on the location, altitude, and yield of nuclear explosions.

Geodetic satellites are used to produce accurate maps of the Earth using photographic and radar techniques, and to provide data about the Earth's gravitational and magnetic fields. Among other things, this information enables the trajectories of ballistic missiles to be planned precisely and also yields the mapping data used in navigation systems on aircraft and cruise missiles.

Surveillance and Reconnaissance

Surveillance and reconnaissance satellites, also known as spy or intelligence-gathering satellites, are used to obtain information about a multitude of military activities.[137] Surveillance is a relatively regular monitoring activity whereas reconnaissance is a search for intelligence of a more urgent nature. Although the tasks are different, they are increasingly combined on a single platform equipped with several sensors.

Photographic spy satellites provide images of objects and areas of interest using optical, infrared, or radar sensors. The information produced can be of incalculable value. The strengths and locations of forces can be determined with considerable precision and weapons under construction or testing can be observed in remarkable detail. It is possible, for instance, to determine the size of an aircraft engine's air intake or the caliber of a tank gun.

The best known American spy satellite is probably the KH-9 "Big Bir" which can perform both wide-area surveillance and "close-look", high resolution reconnaissance with multi-spectral sensors. Big Bird orbits as low as 160 kilometers and is equipped with engines to provide periodic "nudges" to sustain the orbital life for about 200 days. Without these engines, atmospheric drag would normally cause reentry in about a week.

Big Bird's cameras can identify objects as small as 30 centimeters across, its film being processed on board, scanned by an optical system and transmitted to receiving stations on Earth. These images are of poorer quality than photographs, so if exceptionally fine detail is required, Big Bird can jettison film capsules which are recovered in midair by specially equipped HC-130 aircraft based in Hawaii.

The United States also operates the KH-8 photographic satellite which is smaller than Big Bird and also uses the film return technique. This is probably used to photograph intelligence targets of high priority and to supplement other spy satellites.

Big Bird is being phased out in favor of the KH-11 which, instead of using a film system, employs a digital imaging device which reportedly can achieve a higher resolution even from KH-11's higher operational altitude. KH-11 has an orbital life of about two years and its imaging system and ability to transmit data via military communication satellites makes information available almost instantaneously.

KH-12 is soon to be available and will possess a radar imaging system capable of penetrating cloud cover in addition to higher resolution optical and infrared sensors.

Electronic intelligence (ELINT) gathering satellites—also known as ferrets—provide much useful data. They can locate and identify radio and radar transmitters, eavesdrop on communications, and monitor telemetry from missile tests. A radar's purpose can be established by examining pulse width, pulse repetition frequency, transmitter frequency and modulation. As well as aiding with the design of countermeasures, this type of information facilitates operational planning. Known radar and communications sites can be attacked and SAM radars can be avoided for instance.

The United States operates at least two types of ELINT satellites. One is a sub-satellite ferret usually launched along with BIG BIRD and boosted to a circular orbit at an altitude of about 500 kilometers. Another type is the geostationary Rhyolite which bristles with antennae, the largest of which is over 20 meters in diameter. Rhyolite is soon to be replaced by a more sophisticated and highly classified successor possibly known as "Aquacade".

For ocean surveillance and reconnaissance, the United States uses "White Cloud" consisting of three clusters of satellites spaced at 120 degree intervals around a 1,100 kilometer orbit inclined at 65.3 degrees. Each cluster consists of three small satellites dispersed around one "parent" satellite. The clusters orbit in a formation which allows the data from infrared and millimeter wave sensors on each satellite to be collated, providing information on surface vessels more than 3,000 kilometers away. Planned follow-on systems as well as having finer resolution may also employ radar sensors.

Other NATO nations are also developing intelligence-gathering satellites. In the early 1990s, France will commence operations with the first non-American Western intelligence-gathering satellite. This satellite, known as Helios, will be built and operated jointly with Italy and Spain. Helios will reputedly have a resolution of better than one meter and will be largely based on France's SPOT remote sensing satellite. Current plans call for two optical Helios satellites and a third, infrared version. In the mid-1990s, an ELINT model will be launched, followed in the late 1990s by a radar imaging version.[138]

The United Kingdom recently cancelled plans for a Rhyolite-type ELINT satellite known as Zircon and there is speculation that a more advanced type is being planned.

Communications

Satellites are used extensively for military communications: over 70 percent of all American overseas military communications are relayed by satellite, for instance. As the pace and complexity of modern military operations increases, long-range, secure communications become even more essential and this is reflected in the emphasis the military places on communications satellites.

The United States is currently upgrading its Defense Satellite Communications System (DSCS) by replacing DSCS-II satellites with new DSCS-IIIs which have a greater capacity and improved survivability features such as ECM resistance, circuitry protected against electromagnetic pulse (EMP), and propulsion systems to maneuver in order to thwart an attack by anti-satellite interceptors. Similar features also appear on the Fleet Satellite Communication System (FLTSATCOM) satellites due to become fully operational in the early 1990s. Like DSCS, FLTSATCOM employs geostationary satellites for global coverage, but polar regions are poorly served by satellites in geostationary orbits. To plug this gap, a three-satellite network known as the Satellite Data System (SDS) is in operation. This uses highly elliptical orbits and is particularly important for relaying data from intelligence-gathering satellites. Also in operation is the Air Force Satellite Communications System (AFSATCOM) which is not a satellite system in its own right but consists of communications packages carried on satellites such as FLTSATCOM, SDS, DSCS, and others.

From the early 1990s all these systems will be supplemented and eventually superseded by the Military Strategic Tactical and Relay (MILSTAR) system. This will use millimeter-wave region transmissions which provide an enormous communications capacity. Four satellites plus one in-orbit spare will be in geostationary orbits and three will be in highly elliptical polar orbits. An unspecified number of spares will also be placed in supersynchronous orbits at altitudes of 177,000 kilometers (almost half-way to the Moon). MILSTARS will be hardened against lasers and EMP and will also be cross-linked by laser and other communications systems.

France also operates a spaced-based military communications system known as Syracuse 1. This consists of military communications packages fixed to France's civilian Telecom 1 satellites and is used by French forces abroad. In 1991, Syracuse 2 will commence operations in conjunction with Telecom 2 satellites. Syracuse 2 will have a higher capacity and greater resistance to jamming than Syracuse 1.

The United Kingdom was the first nation to operate a geostationary military communications satellite although the satellite itself—Skynet—

was American built. Various Skynet satellites have been launched since the first in November 1969, and two new Skynet 4 satellites are scheduled for launch around 1990. These will support strategic, maritime, and tactical land operations and will be resistant to EMP and jamming.

NATO also operates communications satellites designated NATO III A, B, and C, and a fourth—NATO III D—is an in-orbit spare. These are fully interoperable with DSCS III and Skynet 4, and provide secure communications links between national command authorities and NATO's command structure. In 1990, two NATO IV satellites are scheduled for launch to meet anticipated increases in communications requirements. These satellites will be largely derived from Skynet 4.

Navigation

The space system likely to have the most profound impact on conventional terrestrial military operations is the Global Positioning System (GPS). Already partially deployed, the full system should be complete by 1990. As currently planned, GPS will consist of 21 Navstar satellites placed at regular intervals around three orbital "rings" each inclined at 63 degrees. Orbital altitude will be about 20,000 kilometers so that each satellite orbits the Earth every 12 hours. The satellites will transmit two codes, one for military use and the other for civilian use. The former will enable a user equipped with a small receiver to establish position in three dimensions to within about 10 meters and velocity to within a few centimeters per second. The military signal will be encrypted and highly resistant to jamming. With the 21-satellite network, some regions will experience one or two short periods—less than half an hour—each day when accuracy will be degraded. Funds have therefore been requested for three additional Navstars so that a 24-satellite network with no coverage "gaps" could be in place by 1995.[139]

The applications of GPS are virtually limitless. Ships, aircraft, missiles, vehicles, and even individual combatants will be able to establish position with unprecedented accuracy using passive receivers. Weapons will therefore be able to find their targets with remarkable precision over any range. Submarine launched ballistic missiles, for instance, will be able to achieve the same accuracy as their land-based counterparts; cruise missiles and aircraft will be able to navigate without emitting any signals; combined operations will be easier to manage; minesweeping will be more efficient since swept areas will be mapped precisely and vessels will be able to navigate accurately through them; aircraft will be able to rendez-vous for in-flight refuelling without radar or radio assistance; indirect fire will be better coordinated; and reconnaissance satellites will be able to provide more accurate information about potential targets.

Meteorology

The only Western meteorological satellites used exclusively for military purposes are the American Meteorological Satellite Program (DMSP) satellites, two of which are kept operational. The sensor suite on these satellites is very comprehensive including optical and infrared systems and devices which measure the vertical temperature and water vapour content in the atmosphere. Other sensors provide data on auroral and ionospheric conditions.

This information can be obtained by vessels at sea and by tactical commanders using small, mobile receiving stations. The information gathered is of enormous value. Strategic and tactical air missions can be planned to take account of prevailing weather conditions and missile guidance systems can be up-dated with meteorological data so that in aiming re-entry vehicles compensation can be made for weather conditions. Knowledge of the upper and lower atmosphere is also essential for the efficient planning of satellite reconnaissance missions, and ionospheric data allows the quality of high frequency radio activity to be predicted.

Future Systems

Research is in progress to develop satellites for new military missions such as communications via laser with submarines and detection of air breathing systems such as bombers and cruise missiles. As regards the latter, the United States plans to launch its "Teal Ruby" satellite around 1990. This will be fitted with extremely sensitive infrared sensors and will investigate the feasibility of detecting and tracking air-breathing systems from space.

The Soviet Union employs an essentially similar array of military satellites and has also developed an anti-satellite interceptor to attack Western systems. The United States has responded by developing an anti-satellite weapon which, unlike the Soviet, rocket-launched device, is fired from a slightly modified F-15 fighter dubbed the "Celestial Eagle". The future of this program is in doubt, however, in view of escalating costs and a Congressional moratorium on testing against targets in space.

Work is in hand on a completely new class of space vehicle, however, which may eventually be used for anti-satellite operations and other military missions. This vehicle, designated the X-30, is essentially an aircraft capable of taking off from an airfield and climbing into orbit. A $600 million technology demonstration program is in progress which will allow a decision in 1989 on whether to embark on construction of the X-30 by the early 1990s at a total cost of about $3.3 billion. This could lead to development of a commercial version used for hypersonic air travel, a hypersonic military reconnaissance/bomber capable of achieving speeds of about Mach 12 on the fringes of space, and a fully re-usable spacecraft for placing payloads in orbit, retrieving satellites, and inspecting—and perhaps destroying—an adversary's space systems.[140]

Conclusion

There is clearly no shortage of innovation in aerospace technology. Advances in aerodynamics, materials, propulsion, and avionics are creating revolutionary opportunities for aerospace systems. Stealth, active control technology, extreme agility, sustained supersonic performance and other developments are facilitating the development of aircraft of unprecedented capability such as the ATF, the Osprey, and—further into the future—the X-wing and the aerospaceplane. Complementing these vehicles are weapons systems which are smarter, more agile, longer ranged, and more reliable.

In the past, however, state-of-the-art aircraft and missiles have been difficult to maintain, relatively unreliable, and even more expensive than anticipated. The source of many—though admittedly not all—of these problems has been the electronics systems upon which such systems rely. Consequently, it is not surprising that some reservations are expressed about the chances of avoiding such problems in the future with new systems even more heavily dependent upon electronics. Reports about developments such as VHSIC are certainly encouraging even though similar technological optimism on previous occasions has proven disappointingly ill-founded. More encouraging therefore is the evident recognition of past problems and the resultant insistence on according reliability and ease of maintenance the same priority as other, more traditional performance parameters. Refusal to accept equipment which fails to meet such rigorous standards is worth the inconvenience in delayed deployment.

In view of the critical role that electronics will play in future aircraft, it is worth pointing out that some recently deployed systems have performed exceptionally well. Active control technology—whereby an aircraft flies exceptionaly well using computers and cannot fly at all without them—has proven to be reliable on aircraft such as the F-16 with no aircraft being lost through control failure.

Indeed, commercial aircraft now in production use that same technology. Even though active control systems typically employ quadruple redundancy, this current performance bodes well for the future.

As regards space systems, the trend towards more Western nations using space systems is a positive development for both military and political reasons. A succession of launcher failures in 1985 and 1986 caused many problems for the American military and showed the need for redundancy in military space systems. In political terms, photographs from commercial remote sensing satellites have already added credibility to various American claims about Soviet military capabilities. The United States, for instance, drew attention to the Soviet build-up in the Kola Peninsula and to Soviet directed energy facilities in Dushanbe near the Afghan border but as a matter of policy did not reveal its sources. Subsequent analysis of photographs produced by France's SPOT satellites confirmed these facts.

Space systems for communication, intelligence gathering, navigation, meteorology, and other purposes are enormously useful. Such systems are force multipliers which considerably enhance the effectiveness of terrestrial military operations. It is, of course, dangerous to rely exclusively on such systems but it is certainly worthwhile exploring and exploiting the opportunities they offer. The Global Positioning System, for instance, is likely to find a host of applications beyond those currently projected as experience with it grows in the coming decade.

In sum, provided suitable attention is paid to reliability, ease of maintenance, and cost control, new aerospace technology will provide many of the means to maintain and improve the effectiveness of Western military forces.

5

NAVAL TECHNOLOGY

Technology has revolutionized naval warfare since the Second World War so that today the technological trends are towards consolidation—ensuring that available technologies are fully exploited—and incremental improvement. This chapter charts recent and planned advances in naval technology and the ways these advances affect naval vessels and naval weapons systems. Surface vessels and the weapons systems used by and against them are examined first, followed by sub-surface vessels and then mine warfare. Since naval forces are essentially multipurpose in nature and essentially similar weapons can be mounted on airborne, surface, and sub-surface platforms, some overlap in the various sections is inevitable.

Surface Vessels

Surface vessels are a design compromise. They face threats from other surface vessels, aircraft, submarines, and shore-based systems. Ideally, they should be able to pose a counter threat against each of these. Unfortunately, it is not possible to endow all surface vessels with all these functions so individual vessels are optimized for particular tasks and operate in groups which possess an appropriate balance of capabilities. Even so, some general trends in warship design are evident.

In general terms, the most obvious of these is the trend towards increasing use of electronic systems. Modern warfare requires evermore capable sensors, communications, data processing, and automated systems. Warships already incorporate several miles of wiring and many tons of electronic equipment to handle these tasks. In future, greater use will be made of data buses, fiber optic cables, and more advanced, more compact electronics. Another trend is towards decreasing manpower, the single most expensive item in running a navy. By enhancing reliability, built-in redundancy, repair-by-replacement, and automatic fault diagnosis and repair, it should be possible to reduce manning levels substantially.[141] The latest British anti-submarine warfare frigate, the Type 23, will require a crew of only 140 compared with 200 on its predecessor. Studies suggest that the next generation vessel could be crewed by about 50 seamen.

A further trend is the shift away from missile launchers, which require repeated re-loading and must be trained towards their targets, in favor of vertical launching systems. Most types of missile lend themselves to vertical launch with little modification, and this launch method avoids all the electro-mechanical problems associated with transporting heavy missiles from a magazine below deck. In addition, response time is increased, an important factor in dealing with multiple engagements.[142] Virtually all American destroyers are to be fitted with vertical launch systems over the next 10 to 15 years and future British and French warships will increasingly employ such systems. Other trends are towards smaller radar cross sections and increased propulsion efficiency.

Organic Air Power

As regards the overall composition of surface fleets, a striking trend has been the decline of the traditional aircraft carrier. Within NATO, only the United States and France still operate carriers with conventional take-off and landing (CTOL) aircraft while the United Kingdom, Spain and Italy operate carriers with vertical/short take-off and landing (V/STOL) aircraft. Equally striking has been the trend towards equipping most vessels of frigate size and above with one or even two helicopters. Although no substitute for fixed-wing aircraft, helicopters greatly enhance anti-submarine warfare capabilities, can furnish over-the-horizon targeting information for ship-launched anti-ship missiles, and can even prosecute anti-ship attacks using air-launched anti-ship missiles.

In future, the number of CTOL carriers will increase as the United States adds to inventory and France replaces its conventionally-powered carriers with two nuclear-powered models. V/STOL carrier inventories are likely to remain static since none are more than ten years old. The inventories of rotary-wing aircraft on other vessels will no doubt increase because such aircraft add so much to overall capability. Furthermore, the performance of rotary wing aircraft is set to rise dramatically as the novel systems now under development reach maturity. As discussed earlier, the United States is currently working on two types of rotary wing aircraft which combine the characteristics of the helicopter and the fixed-wing aircraft. One of these, the V-22 Osprey, has a large rotor at the end of each wing which can tilt through 90 degrees to permit vertical take-off and landing, and hovering but which act as traditional propellers for forward flight. This is scheduled to enter service in the early 1990s. The other, the "X-wing", has a stubby four-bladed rotor which is static in forward flight and acts as a wing. This project is still in the research stage but offers all the flexibility of a helicopter plus very high speed forward flight. Vehicles based on this concept are likely to find their way into naval and other inventories but probably not until around the turn of the century.

Another development is the British Skyhook which can be fitted to vessels of 5,000 tons and above. Skyhook is a stabilized crane which will lift V/STOL aircraft such as the British Harrier outboard from a ship's deck. Skyhook then automatically disengages when it senses upward aircraft movement. For recovery, the aircraft matches speed alongside the host vessel in a notional three meter cube and Skyhook latches on automatically. Refuelling can also take place through Skyhook without full deck recovery. This system could be fitted to existing vessels to provide the option of operating V/STOL aircraft instead of helicopters. Another possibility is to build a small escort carrier of about 5,000 tons equipped with two Skyhooks, five or more Harriers, one or more helicopters, and any other mix of weapons systems deemed appropriate. At present, there are no firm plans for using Skyhook but given the expense and scarcity of organic fixed-wing capability using dedicated carriers, and the considerable utility of fixed-wing assets, the system has a good chance of appearing in naval inventories.[143]

Anti-Ship Missiles and Long-Range Targeting

One of the most visible changes in surface vessels since the Second World War has been the virtual disappearance of the heavy naval gun. This weapon, once the primary surface-to-surface naval armament, only exists on four recommissioned American battleships and is intended mainly for coastal bombardment. Nowadays the naval gun tends to be of medium caliber, highly automated, and intended only as secondary armament or for supporting forces ashore. The principal surface-to-surface weapon is now the anti-ship missile.

Anti-ship missiles come in many shapes and sizes each with particular characteristics. Most current Western types are sub-sonic surface skimmers which home in using infrared or radar seekers. Probably the best known is the French Exocet, which has a range of about 70 kilometers and which can be launched by surface vessels, submarines, aircraft, and land-based launchers. Such missiles can only be detected by ship-borne radar about one minute before impact, giving defensive systems little time to react.

Exocet, however, is by no means the most capable anti-ship missile in the West. The Norwegian Penguin, the British Sea Eagle, and the American Harpoon and Tomahawk offer equivalent or better performance. The Tomahawk, for instance, has a range of about 675 kilometers and although rather larger than Exocet has a smaller radar cross section. Sea Eagle can also be programd so that sequentially launched missiles will arrive simultaneously at their targets from different directions.[144]

The trends in anti-ship missiles are towards higher speeds, longer range, greater resistance to counter-measures, and increasing sophistication. Higher speeds clearly complicate defenses since warning times become shorter while longer range increases the security of the launch platform. Greater resistance to counter-measures (dual-mode sensors, electronic counter-counter-measures, etc.) obviously improves weapon lethality, and increasing sophistication (allowing mid-course up-dates of target information, deceptive target approaches, autonomous target search patterns, etc.) improves weapon flexibility.

The current drawback for ship launched anti-ship missiles is that their range exceeds that of ship-borne sensors so considerable attention is now being paid to extending surface vessels' surveillance capabilities in order to make full use of anti-ship missiles and also to provide greater warning of threats such as aircraft and surface vessels and incoming missiles. Ship-borne helicopters are already used to carry radars aloft in order to extend the radar horizon, but because such helicopters have other tasks to perform—notably anti-submarine warfare—other options are being investigated.

One approach is to use over-the-horizon radar which bounces signals off reflective layers in the upper atmosphere. Present systems are too large for deployment aboard surface vessels but progress towards shipboard systems is being made. A more immediately promising approach receiving considerable attention is the use of remotely piloted vehicles. The United States Navy is testing Israeli-built Pioneer RPVs aboard recommissioned battleships and in December 1986 the NATO naval armaments group began a two-year study of ship-borne RPV potential on vessels of 500 to 8,000 tons.[145] Vertical take-off and landing RPVs are considered the prime candidates for the mission due to the space constraints involved. RPVs are also being studied as possible data relays, electronic warfare platforms, and decoys.

Another novel approach being investigated by the United States is the use of airships as radar platforms. The first prototype, the Sentinel 5,000, was to have made its first flight by the end of 1990 and up to five further prototypes were planned leading to possible deployment of up to 50 systems by the end of the century. Funding cuts in 1988, however, led to the deletion of the program from the United States Navy's budget but the Defense Advanced Research Project Agency plus private funding may keep the project going until the concept is validated.[146] By then, the Navy may have both a clear requirement for such a system and the funds to acquire it. If ultimately constructed, the Sentinel 5,000 which is larger than most frigates, will carry a huge radar inside its envelope. As well as the advantage of large radar size, Sentinel 5,000 will be able to remain aloft for much longer than traditional aircraft. Land-based aircraft typically remain airborne for up to 18 hours while helicopters can manage only three or four. Sentinel 5,000 will be able to operate for up to three days without re-fuelling and will be re-fuelled at sea by surface vessels. Cruising speed will be about 40 knots and maximum speed about 80 knots. Operational altitude will be between 1,500 and 3,000 meters. To reduce its radar signature, extensive use will be made of composite materials in the fins and gondola.[147] Although surveillance is the primary mission currently envisaged for airships, other roles such as minesweeping are also considered future possibilities.

Surveillance airships, if they live up to their promise, will complement the airborne surveillance assets of carrier battle groups, provide organic airborne surveillance for surface groups currently lacking such a capability and operate independently, patrolling areas and calling up shore-based aircraft or naval forces as necessary.

Surface vessels do not rely exclusively on organic surveillance assets, however. Land-based radar, ocean surveillance satellites, land-based maritime patrol aircraft, other naval units, and surveillance systems primarily devoted to submarine detection are also used. Airborne radar systems are being refined to provide remarkably detailed surveillance data. It is now possible, for instance, to assemble a radar picture of a ship's silhouette by analyzing Doppler shifts in signal echoes caused by the target vessel's pitching, rolling, and yawing. This new capability is being added to radars on some carrier-based and land-based surveillance aircraft.[148]

Defensive Armament

Ocean surveillance is needed for defensive as well as offensive operations. The Soviet Union has an extremely large inventory of anti-ship missiles, many of which are nuclear armed on surface, subsurface, and aerial platforms. These missiles are generally much faster and larger than their Western counterparts and their range is equally impressive. The air-launched AS-4, for instance, has a range of about 450 kilometers at Mach 4, and the surface- or submarine-launched SS-N-19 has a range of about 350 kilometers at Mach 2.5. Most current models fly relatively high and dive down onto their targets, although recent reports indicate increasing Soviet interest in surface-skimming trajectories. Some types can also be programd so that salvos arrive at their target simultaneously.[149]

The first line of defense against such missiles is, of course, to attack the launch platform, but success is by no means assured. Even a well-defended carrier battle group cannot rely upon eliminating every threatening submarine, surface vessel, and aircraft before they have a chance to launch their missiles, and naval forces without a carrier's air assets would be even harder pressed. Furthermore, since the early 1970s the Soviet Union has practised using satellites to obtain targeting data which, even allowing for delays in analysis and distribution, is probably accurate enough for a missile's sensors to pick up their intended targets during flight. As a result, Western naval forces must employ defensive systems to deal with the anti-ship missile threat. Such defenses include defensive missiles (which can also attack aircraft); rapid-firing, radar-controlled guns, chaff mortars and rockets; decoys such as flares; and a variety of electronic counter-measures. Associated with these defenses are many types of radar, communications, and data processing equipment which must be able to react in seconds. Indeed, in conflicts such as the Falklands War, the first warning of attack noticed by vessels' crews was often the automatic launch of a defensive system.

One of the best examples of such sophisticated defensive systems is the American Aegis cruiser whose primary role is defense against aircraft and missiles. Using a highly capable phased array radar system, Aegis vessels can engage air threats at ranges between 6.5 and 130 kilometers and at altitudes up to 24,000 meters using vertically-launched Standard surface-to-air missiles. For "last ditch" defenses, the British Sea Wolf system probably represents the current pinnacle of missile effectiveness while the American Phalanx, the Dutch Goalkeeper, and the Spanish Meroka—all rapid-fire guns—reflect an effective alternative approach. All these systems are relatively recent developments so the future trend is towards incremental improvement.

As regards the more distant future, lasers may eventually be used for close-in defense most likely on large, nuclear powered vessels, but a more conventional technique under investigation is the hyper-velocity missile. Unlike current defensive missiles which travel at around Mach 2, these would travel at about Mach 4, reducing engagement times accordingly.[150]

Sub-Surface Vessels

Both East and West place considerable emphasis on submarine forces. Submarines are extremely difficult to detect and can be equipped with weapons suited to a variety of missions. Vessels fitted with strategic nuclear missiles, for instance, provide extremely survivable platforms for nuclear deterrent forces while others, fitted with conventional weaponry, pose a major threat to hostile maritime assets.

Submarine technology was revolutionized by the advent of nuclear propulsion which enables a vessel to remain submerged for the entire duration of a patrol since nuclear reactors require no oxygen but provide power to extract it from sea water to meet the crew's needs. Within NATO, the United States, the United Kingdom, and France operate nuclear-powered submarines configured for nuclear retaliation and for more traditional naval missions. Canada has also decided to acquire nuclear-powered attack submarines because they are particularly well suited to under-ice operations.

Nuclear-powered submarines are expensive, however, and diesel-electric submarines remain extremely viable vessels; indeed, when operating on batteries they can be quieter and therefore harder to detect than nuclear-powered submarines. Consequently, many Western navies operate diesel-electric boats alone or in addition to nuclear vessels and only the United States has opted for a purely nuclear-powered force.

The principal method for detecting a submerged vessel is to listen for the inevitable noise generated by its internal machinery, propellers, and water flowing past its hull. Active sonar—like radar but using sound instead of radio waves—is also used but has the drawback of revealing the hunter's existence and location. In fact, again like radar, sonar enables the target to locate the emitter at greater ranges than the emitter can locate the target.

To thwart listening devices, power plants are placed on insulated "rafts" to isolate vibration from the hull, and some nuclear power plants when operating at low power are cooled by convection currents to reduce pumping noise. Propellers are designed to minimize noise or, as on new British SSNs, ducted propellers are used. Sound-absorbing coats in layers up to 10 centimeters thick are also fitted on the latest submarines in order to absorb sonar signals and muffle submarine noise.[151]

The challenge for conventional submarine design is to extend underwater operations. Diesel-electric boats must surface or rise to snorkel depth periodically to recharge their batteries, a process which renders them vulnerable to detection by radar and infrared devices, by "sniffers" which can register their exhausts, and by acoustic sensors which can hear diesel engines more easily than quieter electric motors.

Efforts are therefore being devoted to developing new batteries using materials such as silver and zinc which have higher power capacities and, further into the future, to producing closed-cycle engines which will be able to run for long periods without atmospheric oxygen. Fuel cells which use, for example, hydrogen and oxygen are also being studied. These technologies hold promise but are not yet sufficiently mature to supplant the traditional diesel-electric combination.

The trends for all submarines are towards greater hull strength to facilitate deeper diving, greater speed, and increased automation. Also, and of key importance, data processing capabilities are being improved in order to enhance analysis of data from acoustic sensors built into the hull or towed behind the submarine. This analysis, plus associated functions such as fire control, is extremely complex and requires tons of computer equipment.

In the long term, wholly automated submarines are likely to be used for barrier operations and hazardous operations near enemy home bases. Scicon, a British company, has developed a concept entitled SPUR which would perform these tasks, and similar projects are under investigation in Canada and the United States.[152] Such vessels could be very small and quiet since there would be no crew and associated life support machinery. A fully autonomous submarine is a practical possibility by the turn of the century and might be preceded by semi-autonomous vessels which would receive instructions periodically from shore, surface vessels or aircraft using "burst" radio transmissions.

Submarine Armament

A wide variety of weaponry is available on modern submarines. Unguided torpedoes are still widely used for use against surface targets but guided torpedoes are necessary to engage submarine targets. These use passive or active sensors to home in on their targets and some can be wire-guided from the parent vessel. Ranges are typically about 25 kilometers. A variation on the theme is to extend the distance by fitting a homing torpedo to a missile which carries it to a search area. The American SEA LANCE now under development, for instance, will carry a homing torpedo—or a nuclear depth charge—in excess of 55 kilometers.[153] As regards defenses against homing torpedoes, submarines can conduct evasive manoeuvres and dispense small decoys such as the British "Bandfish" which emit acoustic signals like those emitted by the submarine itself.[154] There is also some prospect that active defenses will be developed in the form of an anti-torpedo torpedo system but such weapons are at an early and highly classified stage.[155]

Anti-ship missiles are also a standard item in a submarine's armory. Exocet, Harpoon, and Tomahawk are all available in submarine-launched variants and are extensively deployed on Western submarines.

The Tomahawk can also be configured for attacking targets on land with either conventional or nuclear warheads. This capability is being exploited by American *Los Angeles* class boats which are being fitted with vertical launch tubes to accommodate 15 Tomahawks of any configuration. Another weapon particularly well suited for submarine use is the mine which submarines can lay covertly to cause havoc for hostile shipping.

Submarine Communications

The stealth and lethality of submarines render them enormously useful but they still have an Achilles heel. The problem is communications.

Radio waves at normal communications frequencies only penetrate a few meters below the sea's surface, so to receive transmissions a submarine must at predetermined intervals rise to a relatively shallow depth and trail an antenna close to the surface. By rising to shallower depths, however, the submarine becomes more detectable.

A partial solution to this problem is the use of extremely low frequency (ELF) transmissions, typically at about 80 Hertz. ELF transmissions penetrate to a depth of several hundred meters so can be received at operational depths, but there are disadvantages including a very low data transmission rate and the need to trail an antenna behind the submarine. The United States operates one ELF system and is building another and the United Kingdom is testing an ELF system. The American system uses three-letter codes which take about five minutes to transmit.[156] These codes can convey much information including the need to rise to shallower depths to receive more detailed communications. ELF systems will liberate submarines from making unnecessary forays into shallower depths but will not dispense with such forays altogether because they cannot cope with all submarine communications requirements. For that—high data rate communications at operational depths—a completely different technology might hold the key: the laser.

Until recently, the blue-green laser was seen as a possible solution to the submarine communications problem. This type of laser beam can penetrate the atmosphere and the sea to considerable depths and experiments showed great promise. The drawback, however was that signals transmitted from satellites or aircraft could strike the submarine at a wide range of angles, too wide in fact for conventional optical receivers at realistic laser power levels. The solution to this problem was the Caesium vapour receiver which is extraordinarily sensitive to deep blue laser light which also propagates well through the atmosphere and the sea. Using this type of receiver it now appears possible to receive laser signals from a satellite-based laser. Consequently, the United States is designing SLCSAT (Submarine Laser Communications Satellite) which, if airborne experiments fulfil their promise, could commence operations in the mid-1990s.[157]

Transmitting signals from submarines is also less than straightforward, always dangerous, but sometimes necessary. Hostile electronic surveillance systems might detect and locate the source of transmissions despite signal compression and highly directional transmission. Consequently, signal buoys such as the American AN/BRT-l are used. These contain a recorded message which is transmitted after a preselected delay to give the submarine time to move away.[158]

Anti-Submarine Warfare

The Soviet submarine threat is one of the greatest challenges to NATO. Soviet attack submarines could severely disrupt NATO's sea lines of communications and ballistic missile carrying submarines hold entire populations at risk. The Soviet Union, of course, faces reciprocial threats from NATO submarines. Not surprisingly, both East and West therefore place a great deal of emphasis on anti-submarine warfare (ASW).

Submarine Detection

ASW is, however, extraordinarily difficult because the ocean is a very large hiding place and submarines can only be detected by acoustic or magnetic methods, both of which have severe limitations. As regards magnetic techniques, submarines—in most cases—are large ferrous objects which inevitably interact with the Earth's magnetic field. The anomalies created in this field can be detected by magnetometers but only over very short ranges, typically less than one kilometer.[159] Consequently, magnetic anomaly detectors are used by submarine-hunting aircraft mainly to establish the precise location of a submarine whose general whereabouts have been determined by other means. This technique can be thwarted to some extent by demagnetizing the submarine or by constructing the hull from a nonferrous metal such as titanium. Neither technique is simple but both are used. The Soviet Alphas are made of titanium, which is expensive and must be welded in oxygen-free surroundings, but which is stronger than steel so the vessel has the additional advantage of being deeper diving.

Efforts are being devoted to improving the sensitivity of magnetic anomaly detectors and advances in high-temperature superconductors could eventually enhance performance dramatically but, for the present, magnetic techniques remain largely as complements to acoustic methods.

Acoustic detection can be startlingly impressive under ideal circumstances. For instance, a Soviet Alpha-class SSN has been picked up in the Norwegian Sea by acoustic sensors as far away as Bermuda.[160] In that case, the submarine was making a lot of noise. Usually, conditions are far from ideal and if a submarine is determined to remain hidden, it can be almost impossible to detect using passive listening devices and difficult even with active sonar.

The problem for acoustic sensing systems is that the sea is an extremely dynamic, heterogeneous environment. Pressure varies with depth, as do salinity and temperature. The latter do not vary regularly, however, but tend to form layers—haloclines and thermoclines—which can reflect sound and which rise and fall according to local conditions and even time of day. In addition, the irregular topography of the ocean floor, surface currents, deep ocean currents, tides, waves, biological activity, and man-made noise sources such as commercial shipping and ocean drilling all have consequences for acoustic detection. Furthermore, in northern areas ice fields generate noise, reflect sounds unpredictably and affect salinity and temperature gradients. These factors either affect the propagation of sound or produce background noise which greatly complicates acoustic detection. In fact, interpreting data from acoustic systems requires understanding of these phenomena and a data processing capability to transform data into usable information.

To make matters worse, in recent years the Soviet Union has made substantial advances in submarine quietening techniques such that the latest vessels such as the Akula-class and Sierra-class SSNs are reputedly 96 to 99 percent quieter than vessels produced a decade ago.[161]

A major element in the West's acoustic detection assets is the American Sound Surveillance Under the Sea (SOSUS) system consisting of hydrophone arrays deployed, for instance, across the Greenland-Iceland-United Kingdom (GIUK) gap, from the Aleutian Islands to Japan, and along the East and West coasts of the United States. These arrays are connected to the shore by cable and information from them is centrally processed. SOSUS is being up-graded to facilitate more rapid analysis and dissemination of submarine information, and is also being supplemented by the Fixed Distribution System (FDS), also known as Ariadne. Ariadne will consist of relatively short range, ultra-sensitive sensors which detect submarine sounds coming in on "reliable acoustic paths" (RAP), loosely analogous to line-of-sight, whereas SOSUS sensors are longer range to detect sounds reflected from the ocean floor, thermoclines, haloclines etc. Ariadne sensors will be deployed at about 10 kilometer intervals in the same regions as SOSUS arrays and will have detection ranges of about 20 kilometers. Unlike SOSUS links which use traditional cable, Ariadne will employ fiber optic links.[162]

Fixed systems are extremely useful "tripwires" but are also extremely vulnerable to enemy action. To plug any gaps in fixed systems and to hunt submarines in areas away from fixed systems, the United States is developing the Surveillance Towed Array Sonar System (SURTASS) and the Rapidly Deployable Surveillance System

(RDSS). SURTASS is a large hydrophone array which will be towed behind specialized oceangoing tugs, known as TAGOS, designed for prolonged, low-speed operations. SURTASS entered service in 1984 and the full 12-vessel complement is due to become operational by 1990. Each vessel will conduct 90-day patrols, trawling for data over large ocean areas. RDSS is an acoustic sensor system which can be dropped from an aircraft or covertly emplaced by submarine. Data can be relayed to a patrol aircraft or via satellite to shore stations. RDSS has, incidentally, been plagued by technical and budgetary problems along with efforts to squeeze too much capability into too small a package.[163]

Towed arrays of various types are also fitted to many surface warships and SSNs. The latter are particularly well suited for operating towed arrays and indeed active sonar because they can operate at whatever depth is most likely to provide the best signals.

Towed arrays are enormously useful but do have limitations. At high speeds they are "deafened" by water flowing past them and they only provide a bearing in relation to their axis, so that a contact could be on either side of the array (technically, in fact, a contact could be anywhere on a "cone" extending from the array at the bearing's angle). Knowledge of local conditions and analysis of the contact strength can yield approximate estimates of range and depth but cannot resolve the "which side" question. Consequently, vessels using towed arrays employ "sprint and drift" tactics, moving at low speed to gather data and moving at higher speed to a new location and bearing to gather new data which can then be correlated to provide an accurate target location.

Work is in hand to improve towed arrays to increase their effectiveness. Data processing advances are helping to extract the systematic noise of a submarine from the continuous, but relatively random background, and new, extremely sensitive sensors are being developed. These sensors use laser interferometry to compare two laser signals, one of which travels along a fixed path while the other travels through a fiber optic cable which is distorted by acoustic pressure. Work on this technique is at an early stage but initial demonstrations indicate that the sensors will be highly sensitive and inexpensive.[164] In addition, efforts are being made to triangulate target location using towed arrays as one source and ship-mounted receivers as the other.[165]

For maximum precision, active sonar yields better results than passive sensors although, as noted, it alerts the quarry that the chase is in progress. Surface warships and submarines are equipped with active sonars, and aircraft can drop sonobuoys which may be active or passive.

The Command Active Multibeam Sonobuoy (CAMBS) is one of the most sophisticated types of sonobuoy. This can operate actively or passively and on command from its control aircraft can vary the depth of its sonar.[166] Variable depth sonars (VDS) can also be towed behind a surface vessel or "dunked" beneath the surface by a helicopter.[167]

For submarine sonars, conformal arrays are likely to become widely used. In the United States, the Advanced Conformal Acoustic Sensor System (ACSAS) is being developed for the next-generation SSN, the SSN-21 Seawolf class. This sensor suite will operate in active or passive modes and will probably use sensors made from piezoelectric plastic placed all round the hull. This material generates an electric current in response to acoustic pressure and will facilitate "all round" surveillance.[168]

A technique known as "bistatic" sonar allows a submarine to listen for echoes generated by a surface vessel's sonar or by a depth charge. This does not disclose the presence of the listening submarine but works only over a relatively short range. In future, a more satisfactory arrangement might be the use of small, robot submarines operating at some distance from the parent vessel, reporting contacts via a fiber optic link.

In order to mask active sonar use, future sonar systems will spread the output signal over several frequencies so that it blends in with random background noise but can still be processed to yield a return echo.

Research is also being directed towards non-acoustic sensing methods. Submarines have many effects on their local environment and, in theory at least, some of these offer some prospects for detection. The sea's surface, for instance, is affected by the topography of the ocean floor even when the ocean is very deep. This suggests that a submarine may create a "bulge" or create distinctive ripple patterns at the surface. Radar satellites have so far failed to detect such phenomena but research continues using both radar satellites and shore-based over-the-horizon backscatter radar.

Submarines also disturb marine micro-organisms causing changes in their bioluminescence. In addition, submarines leak contaminants such as lubricants from propeller shafts and create electromagnetic fields as they move through conductive sea water. Nuclear-powered submarines also generate large numbers of neutrinos which are not contained by any shielding. All these possibilities pose substantial problems but are the subject of research.

Anti-Submarine Weapons

The favored method for destroying a submarine is the homing torpedo. With few exceptions the major NATO nations tend to operate their own particular types but all operate on broadly similar principles. During the early stages of the hunt, the torpedo may be wire-guided from the launch vessel. This delays the target's detection of the torpedo and the launch vessel's location is less likely to be betrayed. Terminal homing is performed using active sonar which tends to be of a high frequency to provide clear target resolution. Torpedoes are usually electrically powered but some use closed-cycle engines.

Advances in Soviet submarines have been such that some can outrun certain Western torpedoes such as the American Mark 46 used by ASW aircraft. The size and resilience of Soviet submarines has also called into question the lethality of many Western torpedoes against such targets. Meeting these challenges is neither easy nor cheap. Torpedoes are complex systems, an intrinsic difficulty being, for example, that at high speeds sonar returns can be masked by the noise of water flowing past the torpedo. Problems such as this have caused delays and cost overruns on many Western torpedo programs.

The trends in torpedo designs are towards increased speed and range and more effective homing capabilities. Examples of specific projects are the British Spearfish and Stingray, and the American Advanced Capability Mark 48. The Spearfish will use a gas turbine and a pump jet to achieve speeds in excess of 100 kilometers per hour while the Stingray will employ active and passive guidance systems and will return target data to the launch vessel through its guidance wires. Stingray will switch to active sonar homing automatically as soon as the target appears to have detected an attack.[169]

The Advanced Capability Mark 48 with a speed of about 102 kilometers per hour will be substantially faster than the standard Mark 48. Research is also being conducted into increasing warhead lethality by using shaped charges and efforts are also in progress to improve torpedoes' on-board signal processing and to use interactive fiber optic links between parent vessel and torpedo. A variation on this latter theme is to develop an air-dropped system consisting of a sonobuoy linked by a fiber optic cable to a torpedo with a radio link between the sonobuoy and the launch aircraft.[170]

Anti-submarine torpedoes can also be launched from parent surface vessels or submarines using missiles to carry them to their target's vicinity. The Australian Ikara, for instance, used by several NATO navies, can carry a lightweight homing torpedo about 20 kilometers from its host surface vessel and the French Malafon and American ASROC have similar capabilties. The American Sea Lance now under development will also enable submarines to launch the Mark 50 Advanced Lightweight Torpedo via missile thereby greatly extending its engagement distance. As an alternative payload, Sea Lance will be able to carry a nuclear depth charge, another effective anti-submarine weapon.

Anti-Submarine Aircraft

Fixed-wing and rotary-wing aircraft also are used extensively in ASW. Fixed-wing land-based aircraft include the land-based American P-3 Orion, the British Nimrod and the French Atlantique, and the American carrier-based S-3 Viking. All these are equipped with sonobuoys, magnetic anomaly detectors, and homing torpedoes. More than 500 Orions are now in service but, having first flown in 1959, some are now reaching the end of their useful life. Consequently, the United States plans to produce a new Long-Range Air ASW Capable Aircraft (LRAACA) to enter service in the early 1990s. The airframe has still to be chosen but specifications call for a 2,500 kilometer range (over 600 kilometers more than the Orion) and the capability to monitor 56 sonobuoys simultaneously compared with the Orion's 16.[171]

Most surface vessels carry one or two helicopters whose primary mission is ASW. Such helicopters can be equipped with sonobuoys, "dunking" sonar, homing torpedoes, and nuclear or conventional depth charges. These vehicles greatly enhance a surface vessel's ASW capability by extending the surveillance area and prosecuting attacks on possible targets. Existing ASW helicopters include the American SH-60B Seahawk and the British Sea King. In future such traditional helicopters may be replaced by tilt-rotor vehicles which offer helicopter versatility and fixed-wing range. Such vehicles may also replace the American carrier-based S-3 Viking fixed-wing ASW aircraft.

Mine Warfare

A naval weapon effective against surface and sub-surface vessels is the mine. In recent years, NATO nations have paid increasing attention to dealing with the Soviet Union's formidable mine capacity and also have recognized the need to expand their own mine-laying capability.

Mine technology has come a long way from the early "hedgehog" impact devices. A high degree of intelligence can now be built into a mine so that it will explode only under vessels of certain types or even under a particular vessel. They can also be programd to detonate after a specified number of vessels have passed by.[172] Such mines usually use a combination of pressure, acoustic, and magnetic sensors to facilitate vessel classification and to prevent spurious detonations which can occur if only one sensor is used. Mines can be deposited on the sea bed in shallow water and at greater depths can be tethered to an anchor so that the mine is suspended at the correct depth below the surface. They can be laid by civilian and military surface vessels, submarines and aircraft.

One of the most sophisticated modern mines is the American CAPTOR (Encapsulated Torpedo) intended for use against hostile submarines and able to be emplaced by surface vessel, submarine, or aircraft. CAPTOR consists of a Mark 46 torpedo in a housing equipped with passive acoustic sensors capable of detecting a submarine at a range of about 1,000 meters. The system is programd to ignore surface vessels but on identifying a submarine, switches to active sonar and determines the optimum launch time for the torpedo.[173]

Also in service is the submarine-launched mobile mine (SLMM) intended for use in shallower water than is the CAPTOR, and designed specifically for covert planting. The SLMM is launched from a standard torpedo tube and then moves under its own power into areas too shallow or hazardous for submarine penetration. The United Kingdom and the United States are currently working together on the Advanced Sea Mine program to produce a more sophisticated version of the SLMM concept and work is also in hand in the United States to develop a new deep water mine for use against Soviet submarines.[174]

The Soviet Union is believed to have a stockpile of around 300,000 mines, a potential threat causing serious concern to Western naval forces.[175] The presence of mines or even their possible presence can greatly hinder naval operations. Two techniques are used to negate mines: sweeping and hunting. Sweeping entails towing equipment which mimics a target vessel causing mines to detonate, or which cuts the mines' mooring cabels so that they float to the surface where they can be destroyed by gunfire. The traditional vehicle for towing such equipment is a wooden (or plastic) hulled vessel which has a very small magnetic signature and which generates very little noise and pressure. More recently, particularly in the United States, helicopters are used to tow submerged counter-measures equipment directly, or to tow a buoyant "sled" to which is attached the countermeasures equipment. This latter method frees the helicopter from having to carry aloft the generators and processing equipment needed to operate the acoustic and magnetic signature producing devices. The sled itself is also relatively immune to mine blast effects and the helicopter too remains out of harm's way. Mine sweeping sleds are becoming more powerful and sophisticated in order to meet the more complex mine threat. Even so, since mines can be programd to remain inert until registering several target passes, it is necessary to complement mine sweeping with mine hunting. Again, helicopters can tow underwater platforms carrying sonar equipment in this case to locate mines which can then be neutralized.[176]

Another method is to use a mine hunting surface vessel fitted with sonar and remote-controlled submersibles which can locate and investigate potential mines. One such submersible is the French PAP 104 fitted with television cameras and capable of placing an explosive charge next to a mine.[177] Mine hunting vessels also can tow submerged sonar vehicles and use divers to investigate and destroy any mines discovered. The trend, however, is towards increasing use of remote controlled vehicles of growing sophistication. The off-shore oil industry has been largely responsible for developing extremely advanced vehicles of this type which can be modified for mine-hunting relatively easily. Such vehicles can conduct detailed examinations of objects on or near the sea bed and, as well as depositing explosive charges alongside them, can if desired cut through mooring cables or even retrieve the object using manipulator arms.

Conclusion

Technology continues to affect the design, structure, and operations of naval forces but the key trends are towards evolutionary improvement rather than revolutionary change. This is hardly surprising since the last few decades have seen many radical advances which have affected naval forces. With respect to surface vessels, heavy naval guns have given way to missiles with ranges an order of magnitude greater and armored construction has given way to sophisticated, active defenses. The nature of organic air power has changed with only a few navies operating traditional carriers of fixed-wing aircraft but with most surface vessels operating rotary-wing aircraft, albeit to perform different functions.

At present, there is a clear need to consolidate some of the changes. Anti-ship missile technology, for instance, has outpaced targeting technologies, a mismatch which is driving efforts to develop long-range surveillance systems so that full use can be made of the anti-ship systems now available.

With respect to sub-surface vessels, the advent of nuclear propulsion has provided the means to remain deeply submerged for extensive periods but communications technology has not yet permitted this capability to be fully exploited.

One of the key questions always addressed in examinations of naval technology is the likelihood of radical progress in anti-submarine warfare. For NATO, one of the most worrying threats is the Soviet submarine fleet whose attack vessels jeopardize NATO's sea lines of communication and whose ballistic missile carrying vessels hold Western populations hostage. Anti-submarine warfare systems are undoubtedly becoming more capable but so are Soviet submarines. In any event, the ASW task is so formidable and the expanses of ocean so vast that only a revolutionary advance in ASW could change the situation profoundly. At present, there is no reason to believe that such an advance is in prospect.

Probably the greatest challenge to Western navies today is not the likely attrition of forces in combat but the almost certain depletion of forces in budgetary battles. Western nations are struggling to maintain their fleets in the face of rapidly rising costs. Some notable successes have occurred such as the increase in the American fleet but this is not matched elsewhere. To mitigate the effects of escalating costs from generation to generation, there is now movement towards international cooperation in procuring naval forces. In a relatively novel arrangement, for instance, the British and the Dutch have exchanged maritime propulsion equipment for point defense systems and eight NATO nations are inovlved in the collaborative development of a fgrigate for deployment in the 1990s. This project, the NFR-90 is NATO's first common program to build a major warship. Current plans call for 50 vessels at a cost of about $200 million each. The project has already run into some problems—disputes over displacement and armament for instance—but some form of vessel with common features for all participants who stay involved will probably emerge. In the future, it would certainly not be surprising to see greater international cooperation in procurement at least in major weapons systems if not in complete vessels.

6

NATO AND NEW TECHNOLOGY

NATO faces a bewildering array of new technologies which offer the prospect of substantial and in some cases dramatic improvements in capability. In the past, however, escalating acquisition and operating costs, excessive duplication in procurement, mismatches between Alliance and national priorities, technology transfer problems, and transatlantic asymmetry in procurement structure have caused NATO to underutilize its technological resources. True, technological progress has led to weapons of unprecedented capability but all too often that capability has been more expensive and less reliable than anticipated, as well as being fielded in smaller quantities than expected. The underlying reasons for this have only recently begun to be addressed, and NATO still runs the risk that new weapons systems based on even more sophisticated technologies and linked together in increasingly complex ways will be even more expensive, less reliable and scarcer than their predecessors. What is being done to avoid this? And what more could and should be done?

Harmonizing Research and Development

It is axiomatic that NATO should not waste research and development resources by duplicating efforts. Furthermore, if collaboration and coordination take place early in the procurement cycle it is easier to cooperate later, and in any event R&D is expensive and a worthwhile target for rationalization. Many cooperative R&D programs exist already and the time seems ripe now to suggest that NATO introduce some form of common R&D fund, a notion that seemed almost a heresy until recently. Several leading NATO figures, including Lord Carrington and Netherlands Defense Secretary Jan van Houwelingen, have supported common funding and the North Atlantic Council has directed the Conference of National Armaments Directors to examine the possibility.[178] Many specific options could suffice. One possibility would be to establish a funding structure akin to that of the European Space Agency which has successfully established Europe as a leader in space technology, demonstrating in the process that collective efforts can produce results far beyond the reach of individual ones. According to the ESA formula, participating nations make contributions in two ways: first, an obligatory contribution based on GNP to fund a program of scientific activities, and, second, optional contributions to various applications programs. In both cases participants are assured of a fair return for their national industries. A NATO R&D fund would not be categorized into "science" and "applications" but the concept of funding a combination of mandatory and optional programs with a fair return on contributions could be attractive.

Another possibility would be to reduce the costs of duplicative R&D and provide a vehicle for coordination by adopting a looser formula than the ESA model which requires a degree of centralized program selection and prioritization. In this case the analogy would be the European Economic Community's BRITE project. BRITE (Basic Research in Industrial Technology for Europe) seeks to identify technological areas of interest to industry and research institutions, but which are only worth exploring on a cooperative basis. The BRITE program puts interested parties in touch with each other and, if a joint research project is deemed worthwhile, EEC funds can be added to those of participating groups. In a NATO context, the projects sought would not be those wanting funds, but those being duplicated. These would be identified and would qualify for funding from a common "pool" provided that the separate national efforts were integrated into one common effort. The detailed, specific mechanisms for this procedure would naturally have to be worked out but the essential operating principle would be to extract obligatory contributions from each NATO nation which could only be "bought back" by eliminating duplicative R&D efforts and embarking on cooperative ones.

Other structures of a common R&D fund can be envisaged, each with particular advantages and disadvantages. The crucial point, however, is that some form of common R&D fund be established.

At a European level, support for a common research fund was contained in a report by an independent study team commissioned by the Independent European Program Group (IEPG). The team's proposals included the recommendation "that a proportion of existing national defense research funds be diverted to establish a common budget and that discussions proceed in parallel to prepare an initial common research program"[179]. The team went on to propose that the common research budget should rise eventually to 500 million European Currency Units, and that the European Space Agency should be used as a model for managing this fund. Governmental response to this has been cautious but movement in that direction may yet occur.

Cooperation and Competition in Procurement

Motivation for NATO to improve its procurement methods ought to be provided by analysis of the Warsaw Pact's procurement practices. This demonstrates that central planning can yield remarkable results in the defense sector, even though when extended to the civilian sector centralization is less productive than a decentralized free market. The evidence is clear: the Warsaw Pact spends less than NATO on defense but achieves more. It would be naive to suggest NATO ought to model procurement on Warsaw Pact lines though. The military, political, and economic relationships between NATO nations bear no resemblance to those of the Warsaw Pact nations and would not lend themselves to supra-national authority. Nor would they move easily to the other extreme of purely open defense markets and free competition. Consequently, the Alliance must seek a better blend of cooperation, coordination and open markets compatible with the needs and constraints of a group of 16 sovereign, democratic nations. The current blend fosters a proliferation of similar programs directed towards similar goals so methods must be found to reduce these duplicative efforts without violating perceived national interests.

The adoption of a more judicious blend of enhanced defense cooperation and competition will be difficult to achieve but signs of progress are in evidence. For example, the United States Department of Defense has been instructed to take "full advantage of technologies, including emerging technologies, to improve conventional defenses by actively seeking cooperative armament programs with the Allies"[180]. The Nunn Amendment authorizes $200 million for cooperative research and development projects with the Allies: NATO's Conference of National Armaments Directors (CNAD) has made great progress in harmonizing defense procurement in the Alliance: the IEPG has been reinvigorated and is now a significant forum for enhancing European cooperative efforts; and there is a long list of bilateral and multilateral cooperation programs established on a government-to-government basis. The Pentagon has also been directed to "expedite filling material needs through acquisition of existing equipment from Alliance nations": and the IEPG nations have agreed to exercise (a) "greater discipline in not launching a national development project where one already exists elsewhere, nationally or collaboratively, in Europe; (b) greater readiness to adopt an equipment already in production or, where otherwise acceptable, to adopt a US project for licensed production in Europe, desirably in conjunction with European allies ..."[181]

In addition, NATO's Defense Planning Committee has endorsed a Conceptual Military Framework (CMF) which provides nations "with broad, longer term guidance on the military requirements of NATO strategy".[182] The CMF draws together for the first time threat analysis, net assessment, critical deficiencies in forces structure and planning priorities and goals and it looks further forward—20 years—than previous planning guidelines. The CMF is a welcome, and long overdue step towards coordinated force planning. Inevitably, however, sensitivities over national sovereignty and the consensual approach to policy formulation prevent NATO from drawing up a prescriptive set of priorities and nations will probably be able to use the CMF's broad guidelines as justification for national plans. This might not be the case with the CMF's supporting documents drawn up by the Major NATO Commanders (MNC). These studies, produced by SACEUR, SACLANT and CINCHAN are more specific and contain more rigorious evaluations of and prescriptions for requirements, force goals and potential for exploiting new technologies. If adopted, these could provide a basis for a better coordinated procurement strategy.[183]

Furthermore, within NATO there is movement towards formulating such a strategy, the Conventional Armaments Planning System (CAPS). The basic aim of CAPS is to establish a set of future NATO force requirements called "NATO Armaments Goals" plus a monitoring and review system which will allow various national armaments plans to be measured against perceived NATO requirements. That the NATO nations have been unable to establish such a process before now is perhaps surprising, but that they have now done so on a trial basis is certainly welcome.[184]

Another method of cutting procurement costs is by opening markets to competition. As noted earlier, one estimate suggests unit savings of 20 to 30 percent if NATO became a free trade area in weapons procurement, a figure which seems wholly plausible if recent American and British experiences are anything to go by.

Following the publicity in the United States surrounding some procurement follies such as a $600 hammer and a $2,000 aircraft coffee pot, many formerly sole source contracts were opened to competition of a relatively gentle nature, with orders being split—usually 60/40 or 70/30—between lower and higher bidders. This process has produced some remarkable savings including $3.8 billion over two years on spare parts, four-year savings of $582 million in Sparrow and Sidewinder missile procurement, and a 57 percent cut in the price of the Maverick missile.[185] In the United Kingdom, increased use of competitive tendering has effected substantial reductions in equipment costs. Costs of a new trainer, armored repair and recovery vehicles, and mine counter-measure vessels have been reduced by 35 percent, 12 percent and 20 percent respectively.[186]

With respect to opening markets to foreign competition, progress is being made too. In June 1987, the IEPG Defense Ministers declared their determination "to develop a step-by-step approach to opening up their markets"[187] and in the United States the Quayle Amendment allows European companies to compete with American companies for United States defense contracts on an equal basis.[188] The United Kingdom and France have also agreed to open their respective defense contracts to bids from each others contractors and, incidentally, to identify at an early stage projects for possible collaboration.[189] Certainly there is scope for more competition. At present, only about 25 percent of NATO procurement by value is open to competitive tendering.[190]

Practices which fall somewhere between cooperation and competition are "project packaging" and "cross purchasing". The former essentially entails trade-offs across several projects to provide more opportunities for "carving them up" rationally. According to Robin Beard, the former NATO Assistant Secretary General for Defense Support, project packaging now justifies urgent attention

> "Packaging projects together, or achieving some sort of rolling accounting system across projects, would seek to balance out work and cost sharing agreements and compensatory arrangements over a number of programs, rather than continuing with the present grisly process of negotiating on a project-by-project basis".[191]

Cross purchasing is more of a simple barter whereby nations essentially agree to swap systems they each need. The net effect of this is a division of labor whereby each nation circumvents the additional expense of joint procurement while being assured a larger production run. Such arrangements are difficult to negotiate and raise concerns about dependence upon foreign suppliers and sacrificing national competences, and opportunities are scarce because of different national procurement schedules and requirements. Even so, some successful trades have occurred such as British marine propulsion systems for Dutch point defense systems, and it appears that the United Kingdom and France are actively exploring the possibilities for similar arrangements on a regular basis as are Italy and the Federal Republic of Germany.[192] As NATO moves towards greater coordination of requirements and schedules, more "targets of opportunity" for cross purchasing should emerge and it would be sensible to encourage their exploitation.

Transatlantic Disparities and the European Pillar

Procurement

In 1985, the United States spent approximately $68.3 billion on defense equipment while the European NATO members collectively spent about $17.3 billion. A year earlier, American expenditure on defense research, development, testing and evaluation (RDT&E) was $29.3 billion compared with European expenditure of about $9.2 billion. Furthermore, because European efforts in procurement and RDT&E are fragmented and largely uncoordinated, the real disparities are even greater than these figures suggest. It has been estimated that the costs of European defense procurement are inflated by between 42 and 59 percent as a result of duplicated R&D, inefficient production runs, and the relative paucity of competition for defense contracts.[193]

The funding and structural disparities in European and American procurement give rise to a variety of Alliance difficulties ranging from American complaints—in some quarters—about the burden born by the United States in order to defend Europe, to European fears about American domination of the Alliance defense market. More efficient European defense procurement would at least decrease the intensity of these difficulties and would, at the same time, enhance Europe's and the Alliance's overall defense capability. Consequently, the prescriptions calling for greater harmonization of Alliance R&D and more extensive cooperation and competition in procurement must be applied most vigorously and urgently within Europe. The common European defense R&D fund proposed by the IEPG's defense industry study group therefore seems particularly attractive and many of the group's other proposals have considerable merit. These include the creation of a small, permanent secretariat for the IEPG to identify opportunities for further European collaboration; the stimulation of competition for defense contracts and sub-contracts by, for example, publishing and circulating national contract opportunities throughout Europe; the development of common European technical standards and specifications; and a political commitment to adopt a policy of competition across Europe.[194]

With R&D funding at about one third the American level and procurement funding at only one quarter the American level, Europe already has a substantial handicap. It must therefore reform its approaches to defense R&D and procurement if in future it is to be treated as a more or less equal partner in transatlantic cooperative ventures and be competitive in the American market.

Technology Transfer

Disparities in approaches to technology transfer are another impediment to transatlantic defense cooperation. More extensive American controls on technology exports although intended to curtail the Eastern bloc's access to Western technology also inhibit Allies' access. Progress is being made in resolving the difficulties with several specific improvements in the offing, but still more are needed. A positive development is the movement in both the American Administration and within Congress to remove the need for export licences on sales to COCOM nations and to set a threshold on American content in goods produced abroad below which it would not be necessary to seek American approval for re-export. The threshold level will probably be about 25 to 30 percent so some major problems should be removed. There is also movement towards taking greater account of foreign availability of technology in formulating control lists so that goods which are readily available on the world market would be free from controls. A similar movement is occurring within COCOM, accompanied by efforts to persuade non-COCOM countries to adopt COCOM-type controls. In addition, COCOM nations are increasing the harmonization of national control and enforcement procedures as well as rationalizing control lists to ensure that out-dated controls do not remain in force.

Further improvements could include strengthening penalties for companies which violate export controls, still greater efforts to take account of foreign availability, and greater industrial involvement in the formulation of export controls. Penalties should be harsh not just because violations threaten Western security but also because they would demonstrate to the United States that the Alliance as a whole takes a dim view of control violations. Greater account of foreign availability would enhance the credibility of the control process in the eyes of industry which rightly points out the futility of some controls and therefore questions the validity of others. It should also help to bring the more extensive American control list more into line with COCOM's. Greater industrial involvement in formulating controls would make control more rational while also re-assuring industry that its interests were being considered. Industry tends to have a better appreciation than does government about the state-of-the-art in technology and about its availability so it would be sensible to use this expertise.

As noted earlier, the ideal would be common Alliance export controls so that controls within the Alliance would be unnecessary and technology could be transferred without bureaucratic interference.

Although such a situation is a distant prospect, progress is being made. The question is how long will the reservoir of ill-will created by past problems take to drain? Effective exclusion of American contractors from some European projects and bureaucratic impediments in transatlantic ones, cause extra costs and delays which can only be detrimental to Alliance procurement.[195]

Quality, Quantity, Technology and Doctrine

In broad, almost philosophical terms the NATO nations appear to be embracing a more sophisticated—and long overdue—understanding of the relationship between quality and quantity. The notion of quality and quantity as mutually exclusive parameters, often expressed in terms of "quality versus quantity" seems to be increasingly recognized as an oversimplification of limited utility. Although quality can be traded for quantity and vice versa, this balancing act is by no means inevitable. The concept has, in fact, arisen from implicit acceptance of a definition of quality that has used parameters such as range, speed, height, depth and agility as measures of performance and has taken operational readiness for granted. As a result, NATO has often acquired high technology weaponry whose largely unforeseen, recurring operations and maintenance (O&M) costs have placed a substantial burden on defense budgets.

It is to be hoped that recognition of this problem will lead to its solution but understandable scepticism nevertheless persists. As David Greenwood has written in a discussion of Very Intelligent Surveillance and Target Acquisition (VISTA) technologies and terminally-guided sub-munitions (TGSMs), as "for the propensity of O&M costs to get out of hand, what is striking about the advocacy of VISTA technologies and IGSMs which is now in full spate is the similarity between their proponents' assertions and past claims which have proven spectacularly ill-founded".[196]

If NATO is to relieve pressure on defense budgets and maintain operational capabilities, the parameters of quality must be prioritized differently: reliability and maintainability must rank equally with more traditional yardsticks. Technology can enhance reliability and maintainability so system specifications can and should take account of these requirements.

At the same time, NATO must avoid being seduced by high technology weaponry. In the past, the appeal of such systems combined with underestimates of their acquisition and operational costs has drawn attention away from less glamorous but nevertheless critical areas. As a result, deficiencies have arisen which have been labelled "war stoppers" by no less an authority than General Bernard Rogers, former Supreme Allied Commander Europe. In testimony to the Senate Armed Services Committee, General Rogers listed deficiencies in munitions stocks, spare parts, hardened aircraft shelters, transport aircraft, and chemical weapons which would quickly force NATO to surrender or resort to nuclear weapons within a few days of attack.[197] Progress has been made in rectifying these deficiencies but the situation still well illustrates the opportunity costs of high technology weaponry.

A related shortcoming in procurement strategy is a unit-for-unit replacement philosophy whereby armed services seek to preserve their hallowed and traditional tasks and equipment types instead of pursuing a mission-oriented approach. This process has been described as "doctrinally remote" and has engendered a sterile debate about whether doctrine should lead technology or whether technology should lead doctrine. In fact, doctrine and technology should have a symbiotic relationship. "Outdated doctrine cannot be allowed to constrain technological innovation. But technological salesmanship should not impel adoption of inappropriate concepts of operations".[198] Viable doctrine cannot be formulated without reference to technology's operational limits and potential, nor can technological virtuosity substitute for doctrinal inadequacy.[199]

New, highly versatile technology is now opening up new mission possibilities and methods of conducting old missions in new ways so it is more important than ever to overhaul traditional practices and values. To avoid mismatches between doctrine and technology, procurement decisions should stress the conduct of missions rather than simply the replacement of obsolescent equipment. In selecting new systems, rigorous assessments of new solutions to old problems will be needed to ensure that opportunities are seized and that new technology is applied as appropriate rather than as available.

Technology on the Battlefield

Regardless of the specific systems which will ultimately find their way through the procurement labyrinth, some key trends in the nature of weapons systems are emerging as a result of technological progress.

Delivery Precision

Due to the electronic revolution, sensor and control electronics can now be made sufficiently small to furnish a wide variety of munitions with an advanced guidance capability. For example, lethal RPVs, artillery munitions such as the Copperhead, SADARM, and the Advanced Precision-Guided Munition, land-based missiles such as the MLRS and ATACMS, and air-launched weapons such as HARM, ALARM, Tacit Rainbow, and MSOW all employ sophisticated electronics to enable them to home in on their targets. Widespread use of these types of precision-guided systems should yield several important operational benefits. Most obviously, combat-effectiveness should be improved since lethality will be increased with indirect fire systems in particular becoming much more efficient. Fewer munitions and sorties should be needed to destroy a given number of hostile force elements and destruction should therefore take place more rapidly. Logistic burdens should be reduced and commanders should have greater flexibility in engaging enemy forces as a result of improvements in precision guidance and other factors such as range and mobility.

Range and Mobility

Advances in propulsion technologies, aerodynamics, materials, and electronic design systems are enhancing the range and mobility of both platforms and delivery systems. Platforms such as tanks, helicopters, and fixed-wing aircraft are, in general, becoming faster, longer-ranged and more reliable as are their associated delivery systems. These trends are evident across the full spectrum of weapons from small anti-tank missiles—where systems such as the FOG-M offer substantial range improvements—up to runway-cratering weapons where advances in both the aircraft platforms and the missile delivery vehicles are extending ranges considerably.

Enhancements of range and mobility will provide commanders with a larger number of choices in engaging hostile forces. Targets deep behind the forward line of troops could be engaged to disrupt follow-on forces and airfields, and the capacity to bring forces to bear "sideways" to provide support where necessary will also be increased. Even without wholesale physical displacement, the firepower from disparate force elements such as tube and rocket artillery, RPVs, helicopters and fixed wing aircraft will be able to focus on hostile force concentrations from ever greater distances.

Transferring forces or even just concentrating firepower from forces dispersed over a wider area will require a high degree of coordination. At its most extreme, elements of air, ground, and naval assets could all be called upon to operate side-by-side in an integrated operation whose success would depend on effective use of new systems to perform data acquisition, information processing, and command control and communications.

Force Integration

The increasing precision and range of weapons systems can only be exploited efficiently if information about hostile forces can be gathered, analyzed, and distributed sufficiently rapidly. Information-gathering assets ranging from RPVs, seismic, magnetic, infrared, photo-optical, and acoustic sensors to airborne and space-based systems will provide vast amounts of data on hostile forces which will have to be processed and then re-transmitted to appropriate friendly forces. At the same time, information processing and distribution systems will have to deal with unprecedented amounts of information and commands concerning the status and locations of friendly forces, logistics, and the conduct of operations. To meet these demands, many specific systems are available or under development including battlefield information and/or communications systems such as HEROS, WAVELL, the Position Location Reporting System, the Joint Tactical Information Distribution System, the European Data Distribution System, Ptarmigan, RITA, and—eventually—BICES. Even so, concern remains about the ability of such systems to function reliably in the heat of battle.

This is a critical issue because the successful integration of disparate combat elements into a coordinated defensive fighting force will depend on the performance of these types of systems. If they do prove reliable and of sufficient capacity despite hostile attempts to disrupt them, they will be vital force multipliers; if not, then NATO will fail to achieve the crucial synergy it needs to mount a robust and cohesive defense.

In other words, integrated operations will essentially depend upon the successful execution of a series of steps—in general terms, target location and identification, transfer and analysis of target data, selection of systems to engage the target, transmission of data to friendly systems, and target engagement—and each step may require interaction between several systems. Consequently, the more steps needed to conduct a mission, the more reliable each step must be.

Reliability

The lessons of the past are being applied such that reliability and serviceability are increasingly demanded in specifications whether the item in question is a microchip or an advanced fighter aircraft. This will lead to reduced maintenance burdens and improvements in factors such as availability and sortie rate so that better reliability will be a significant force multiplier. It should also allow movement towards force structures with greater emphasis on the fighting "teeth" and less on the logistic and support "tail". Whether reliability can be improved sufficiently to facilitate some of the more elaborate integrated operations remains an issue of some debate however.

The Nature and Pace of Combat

The "dispersal" of anti-armor capability throughout a wide variety of systems obviously will provide greater flexibility in mounting defensive operations, though the price of this flexibility will be an increase in operational complexity. This flexibility might eventually be reflected in procurement strategies such that less emphasis may be placed on the tank, currently the primary means of blunting an armored assault. As the tank's supremacy becomes increasingly challenged by other anti-armor systems, it would not be surprising to see movement towards these other systems. Abandonment of the tank seems extremely unlikely, however, because it too is being improved and still represents the best means to move into and occupy territory. That ability is critical, even though NATO is a defensive Alliance, in order to launch counter-offensives both to disrupt hostile offensives and to attempt to regain lost ground. Furthermore, a mixture of anti-armor weaponry—including tanks—provides a greater number of defensive possibilities to complicate an opponent's planning. Put another way, removing the threat posed by the tank's combination of mobility and firepower would permit an opponent to devote greater efforts to the offense since he would not need to take substantial precautions against a disruptive ground-based counter attack.

Technological progress should also permit reliance on nuclear weaponry to be reduced, improvements in delivery accuracy and the emergence of, for instance, anti-armor sub-munitions with homing capabilities pose a severe threat to armored formations and, like nuclear systems, could make an aggressor reluctant to concentrate his forces as much as he would otherwise choose. Nuclear systems, incidentally, will also benefit from techniques to enhance range and precision so that new nuclear systems should be more survivable—and therefore less numerous—than their predecessors.

Although increasing weapons precision ought to reduce combat's collateral damage and allow force to be applied more discriminately, for the battle's participants warfare will be no less lethal, and both its pace and intensity will increase markedly. Greater weapon speeds are one obvious reason for this but other factors will be no less significant and even more pervasive. Force availability and sortie rates will increase as reliability improves while greater delivery precision will increase the lethality of each sortie and salvo. Greater range and mobility will enable more systems to bring their firepower to bear on a given area while the more extensive use of all-weather, day and night sensors such as thermal and image-intensifying equipment and airborne radars will reduce or even eliminate lulls in operations due to climatic conditions and nightfall. Thus, the ferocity of battle will be substantially increased by new weapons technologies and the endurance of combatants will be even more severely tested.

Constraints on New Technology

New technology is already opening up scope for new operational concepts. The most obvious manifestation of this is NATO's adoption of "Follow-on Forces Attack" (FOFA) which calls upon air assets to conduct interdiction missions up to several hundred kilometers behind the forward line of troops (FLOT). Technologies and systems relevant to this mission include deep surveillance using equipment such as JSTARS, precise navigation provided by systems like TERPROM, and the plethora of munitions for engaging fixed and mobile targets. Other proposals not yet adopted have called for the development of missiles to conduct offensive counter-air missions—a plan labelled "Counterair 90"—and various emerging technology initiatives which would inject extra funds into an array of new technologies to counter Warsaw Pact forces. At present, NATO is pursuing a more conservative approach which essentially places more emphasis on correcting the less glamorous deficiencies in its force structure—spare parts, munitions stocks, aircraft shelters etc.—while investigating new technologies with a view to deployment when their maturity is more assured.

In considering new concepts, however, NATO is constrained by a multitude of factors which can limit NATO's options considerably. The discussions surrounding FOFA's adoption took place—generally speaking—at two basic levels. The first was the professional debate which embraced issues such as the adequacy of the relevant technology and whether FOFA's requirements would consume procurement resources which could be better devoted to dealing with the threat closer to the forward line of troops. The second was the public debate about the perceived provocative nature of FOFA. In the event, SACEUR's view prevailed but this should not imply that NATO therefore can ignore professional, public, and political criticism. NATO cannot, for instance, override West German reluctance to deploy fixed defenses such as minefields and fortifications which would be seen as formal acknowledgement of the division between West and East Germany. Nor can NATO yet abandon forward defense in West Germany in favor of maneuver warfare because it is deemed politically unacceptable to yield any German soil without a struggle for any reason, including military commonsense.

Political difficulties could also complicate the exploitation of militarily attractive technologies soon to be available. The difficulties in deploying Cruise and Pershing II missiles in several European nations have left a public and political legacy of "missile aversion" which could cause political reluctance to deploy new land-based missiles for years at least. This could mitigate against deployment in Europe of ATACMs or any other missile which however armed could be blighted by the label "nuclear capable". Yet antipathy towards aircraft and new aircraft armament does not seem to be so substantial with the probable consequence that missions which could be conducted more reliably and possibly more cheaply by land-based missiles will be assigned to aircraft which otherwise could be usefully employed elsewhere.

The adoption of new, high technology weaponry may also be constrained by political motives with economic foundations. Initial American enthusiasm for forging ahead with emerging military technologies was met by European scepticism based at least in part on the fear that the United States would dominate the field in such weaponry, relegating Europe to purchaser rather than co-developer and co-producer.

With respect to new technologies, the Alliance will have to take a variety of steps to minimize the role played by considerations beyond the military. The European nations will have to ensure that they keep abreast of technological progress and this reinforces the view expressed earlier that Europe in particular should reduce fragmented and duplicative efforts through greater cooperation and coordination, including the creation of a common research fund. Only if Europe and the United States can participate equitably in the development and production of high technology weaponry will proposals to move in that direction bear substantial fruit.

As regards the deployment of new land-based missiles in Europe, if this is to be attempted, their non-nuclear status should be emphasized, as should their contribution to delaying the need to "go nuclear" in the event of a conflict in Europe. An attempt to deploy such systems in the near future in view of prevailing public and political sentiment seems relatively unlikely, however, so it would be prudent to press ahead as rapidly as possibly with long-range stand-off weapons to ensure that the "deep strike" mission can be conducted effectively.

With respect to West German feelings, it seems that NATO cannot win regarding public acceptance of new concepts. FOFA is regarded with suspicion and alternatives to forward defense are regarded as unacceptable. At some point, however, an extremely difficult choice may have to be faced. Sustaining forward defense credibly may be judged too expensive by those nations outside the Federal Republic of Germany whose forces carry that burden. Should the point come when the choice is between a change in strategy or withdrawal, the former is likely to be preferred. In that event, or preferably sooner, new technology presents many opportunities for posing military obstacles (such as smart mines) and for supporting manoeuvre warfare (such as JSTARS and smart weapons) which would improve NATO's chances of rebuffing a Warsaw Pact attack.

Another factor which is likely to affect decisions on force structures is progress in negotiations on conventional forces deployed in Eastern and Western Europe. The Conventional Stability Talks will certainly focus political and public attention on many issues regarding conventional forces which have in the past been overshadowed by concerns about nuclear weapons. The issues and the technologies relevant to these negotiations will be difficult to explain in popular terms and, indeed, even professionals will be hard-pressed to get to grips with some of the complexities of the subject. Technology in particular is an exceedingly difficult issue if for no other reason than its progress is unpredictable and it seems almost inevitable that once any arms control agreement is reached, technology moves—or is driven—in directions beyond the agreement's scope. Consequently, the flexibility and versatility of modern technology plus the rapid pace of technological innovation will pose a considerable challenge to negotiators ingenuity. On the other hand, technology will offer the opportunity to move towards less threatening force postures, a key goal of the negotiations.

Ideally, a conventional arms limitation agreement should permit each side to maintain the capacity to mount an effective defense while limiting each side's capacity to invade the other. Since the Warsaw Pact clearly possesses a substantial invasion capacity and NATO does not, negotiations will be far from straightforward. The Warsaw Pact will seek reductions in NATO force elements which it finds threatening —notably NATO's tactical air assets and new systems with the ability to strike targets in Warsaw Pact territory—while NATO's first priority will be reductions in Warsaw Pact tank and mobile artillery forces.

There is, however, a profound difference between, on the one hand, forces such as tanks and mobile artillery which are the main means of moving into and holding territory and, on the other hand, forces such as land-based missiles and penetrating aircraft armed with advanced munitions which can disrupt enemy operations deep behind the forward line of troops because the latter—in the absence of a large-scale capability to inject ground forces deep into enemy-held territory—do not represent an invasion threat.

This difference could be difficult to explain to the general public, particularly if negotiations—as is usually the case—are accompanied by self-serving public pronouncements about the issues involved. Since, as mentioned above, some sections of public and political opinion are already likely to be predisposed to be hostile about systems capable of deep strike missions, NATO would be well advised to pre-empt criticism by stressing that the capacity to disrupt enemy activities behind the forward line of troops does not constitute an invasion threat to the Warsaw Pact and that a defensive posture does not preclude the engagement of targets in hostile territory. At the same time, emphasis should be placed on the utility of long-range systems in supporting operations taking place at some distance "laterally" along the forward line of troops.

As regards tanks and mobile artillery, reductions by NATO and the Warsaw Pact should be asymmetric in view of Warsaw Pact superiority in these systems. Unilateral Warsaw Pact cuts are unlikely to be forthcoming, however, so if the Warsaw Pact cannot be persuaded simply to cut its tank forces down to NATO's level—or thereabouts—NATO should be able to place at least some of its tanks on the negotiating table as a result of the growing anti-armor capability of its other force elements. NATO can only move in that direction, however, if other defensive technologies—including those which would permit target engagement at long ranges—are not unduly constrained.

The Northern and Southern Regions

Appraisals of the impact of new technology on military operations generally focus on the Central Region. The Northern and Southern

Regions, however, will be affected by new technology in some specific ways which merit particular attention. In both theatres, for example, the maritime dimension is extremely significant. Indeed, for Norway to be successfully defended, NATO's naval forces must prevail in the Norwegian Sea and, should Norway fall, Soviet access to air bases there would severely—even critically—curtail NATO's naval operations in the Norwegian Sea while also enhancing the threat to NATO's Atlantic sea lines of communications. In the Southern Region too, control of the Mediterranean is vital to success on land. While Soviet naval forces cannot hope to prevail, they could delay and disrupt NATO's sea-borne reinforcements and restrict the use of NATO's naval assets which are needed to support the land battle.

The Northern Region

For the Soviet Union, the Northern Region provides easily the most suitable naval basing area. The Kola Peninsula gives access to the open sea without the hindrances of ice or choke points controlled by potential adversaries. Nor does Kola suffer from the logistic difficulties of the naval base at Petropavlosk on the Kamchatka Peninsula. As a result, around 50 percent of the Soviet sub-surface fleet and about 30 percent of the surface fleet are assigned to the Northern Fleet. Furthermore, over 60 percent of the most modern Soviet SSBNs and almost 70 percent of the latest surface vessels have been added to the Northern Fleet.

The Northern Fleet has, essentially, three tasks. It provides a large part of the Soviet Union's assured nuclear retaliatory capacity, it threatens NATO's sea lines of communications between North America and Western Europe, and—along with Soviet ground and air forces—it threatens northern Norway. The introduction of new technology is clearly enhancing the Soviet Union's ability to perform these tasks but the most profound consequence has been a shift in the concept of operations for SSBNs.

Older SSBNs with relatively short-range missiles have to pass through the surveillance systems in the Greenland-Iceland-United Kingdom (GIUK) gap and must patrol in areas where NATO anti-submarine warfare forces hold sway. Newer SSBNs fitted with longer range missiles such as the SS-N-8, SS-N-18, and SS-N-20, however, can strike their targets from the Barents and Arctic Seas thereby removing the need to run the gauntlet of NATO's anti-submarine warfare forces and instead operating in areas protected by Soviet naval and air forces and under the Arctic ice. Under the Arctic the background noise from shifting ice, changes in salinity and temperature, and the physical ice barrier hamper or even preclude the use of surface and air anti-submarine forces. In the past, Soviet SSBNs would still have had to move into relatively ice-free water to launch their missiles but newer SSBNs can break through several feet of ice so in future they will not have to run the risk of moving to more vulnerable areas to execute missile launches. This may lead to the adoption of an operating technique known as the "ice pick" whereby an SSBN would drift with the ice, practically inert, perhaps for months at a time, ready to forge its way directly to the surface if required.

To counter Soviet forces in the Northern Region, NATO's maritime strategy has already become more "self confident", a development facilitated at least in part by technological progress. Increased emphasis is placed on deploying forces well forward of the GIUK gap. These forces are intended to provide direct support for the land battle, to engage Soviet forces in transit to Norway and the Atlantic, to launch attacks against Soviet bases in the Kola Peninsula, and to hunt for Soviet SSBNs in the Barents and Arctic Seas.

The survivability of NATO's forward presence relies upon both the stealth of nuclear-powered submarines and the defensive capacity of carrier battle groups. The offensive capabilities of these force elements have benefitted considerably from technological advances and by operating well forward can project power in ways which disrupt Soviet offensive efforts and dilute them by compelling Soviet forces to concentrate more on their own defense.

The presence of NATO SSNs in the Norwegian, Barents, and Arctic Seas poses a threat to Soviet surface and sub-surface vessels. This threat consists of submarine-launched anti-ship missiles and torpedoes plus a full spectrum of advanced mines. In addition, American SSNs armed with conventional and nuclear land attack cruise missiles hold at risk Soviet naval installations, air bases and ground forces' facilities. Complicating matters still further, carrier task forces in the Norwegian Sea possess assets which could threaten the Kola Peninsula, support

NATO forces on the ground in Norway, and intercept Soviet submarines, surface vessels, and aircraft acting in support of an offensive against Norway or in transit to the Atlantic.

Sustaining NATO's presence in the Norwegian Sea and beyond would not be cost free. Soviet naval forces and maritime attack aircraft would undoubtedly achieve some successes. Even if NATO surface forces were compelled to retreat, however, Soviet forces would still have to retain defensive capabilities to guard against an SSN threat whose scale would be difficult to appraise due to the uncertainties of anti-submarine warfare operations.

New technology is also enhancing NATO's capability to coordinate its air, sea, and land forces, a capability of particular utility in the Northern Region. Advanced sensor, communications, and data processing systems such as AWACS, and JTIDS will act as force multipliers, facilitating the efficient integrated operation of assets such as ground forces and naval aviation.

Space-based systems too will improve the effectiveness of forces used in the Northern Region. At present, the Northern Region is relatively poorly served by military communications satellites because the efficacy of geostationary satellites decreases as latitude increases. The main "gap filler" is the three-satellite Satellite Data System (SDS) in highly elliptical orbits passing over the Northern Region. SDS is of limited capacity and in time of conflict would be used primarily by the United States Strategic Air Command. In the early 1990s, however, the Milstar system will be operational. Three of these extremely capable satellites will be placed in highly elliptical polar orbits and will greatly improve communications in the Northern Region. The Global Positioning System will also have obvious utility as will new meteorological satellites.

The growing emphasis on forward operations in the Northern Region is somewhat controversial. It is argued that deployment of naval forces so far north could be seen as provocative and that maritime assets could be used as effectively in barrier operations in the GIUK gap and convoy protection in the Atlantic. However, there is no doubt that the presence of NATO naval forces beyond the GIUK gap will force the Soviet Union to conduct defensive operations in the Northern Region thus diminishing its ability to interupt NATO's Atlantic sea lines of communication. Futhermore, it is worth recalling that the battle for the Atlantic would be much more difficult for NATO if Soviet aircraft could operate from airfields in Norway and that the successful defense of Norway requires the successful defense of the Norwegian Sea. Consequently, the defense of the Atlantic sea lines of communication begins in the Northern Region. In addition, in the case of SSN operations against Soviet SSBNs in the Arctic and Barents Seas, these would serve only to maintain the status quo by denying the Soviet Union an SSBN sanctuary. Viewed against this background it is hardly surprising that NATO plans to mount a spirited defense of the Northern Region rather than to cede the initiative to the Soviet Union. As for being provocative, NATO deployments forward of the GIUK gap should be seen as a defensive response to the threat posed by Soviet forces.

The Southern Region

Before examining the impact of new technology on operations in the Southern Region it is interesting to dwell briefly on some of new technology's implications for weapons acquisition in the region. For some nations in the Southern Region—specifically Greece, Portugal, and Turkey—the price of modern weapons is often prohibitively expensive so in large part these countries remain equipped with aged and obsolescent systems. They nevertheless face Warsaw Pact forces which are armed with increasingly advanced weaponry, though not to the same standard as their Central Region counterparts. The more prosperous NATO countries are aware of the equipment problems prevalent in the Southern Region but are unable to donate adequate material. Indeed, they are hard pressed to equip their own forces.

Such donations as do occur can also be less than ideal. Turkey, for instance, has been given four different types of combat aircraft over the years: the F-100, the F-5A, the F-4E, and the F-104G, all of which are still operational. The F-104Gs have been provided by six different nations and are therefore fitted with differing equipment and instruments and have been maintained to differing standards. Not surprisingly, operating such a mixture of aircraft poses substantial difficulties and represents a maintenance burden that Turkey can ill afford.

At first glance, as the major weapons producers within the Alliance move towards systems of still greater sophistication, it appears that—in the absence of dramatic military aid programs—the problems for NATO's most economically disadvantaged nations will become worse. Rising acquisition costs would appear to place new military hardware still further out of reach and maintenance burdens would be even more severe in terms of both financial costs and loss to the economy of much-needed, technically-qualified personnel.

There is hope, however, that in the future acquisition and operational costs will not continue to rise as relentlessly as they have done in the past. The gradual movement towards procurement reform involving enhanced cooperation and competition should limit weapons costs. Weapons will probably not become cheaper because even if research, development and production were wholly rationalized, growing technical sophistication is likely to increase the proportion of R&D in total program costs but the gap between the rate of inflation for defense equipment and the standard rate of inflation might not rise as rapidly as it has done historically. In other words, weapons will continue to be expensive but not as expensive as a projection of current trends would imply.

At the same time, increasing emphasis is being placed on reliability and ease of maintenance so that future weapons should be more "user friendly" in terms of both operations and maintenance. Whereas, in the past, increasing sophistication has required larger numbers of highly skilled operators and maintenance personnel, the tendency now is to build weapons systems which have longer mean times between failures and which require less maintenance time. These trends are becoming evident across the entire spectrum of military equipment from anti-tank weapons to fighter aircraft.

The benefits of increased reliability and ease of maintenance are being realized only gradually but for the less prosperous NATO nations of the Southern Region they cannot arrive too soon. Their advent will render new weaponry more attractive in terms of overall costs and will be less demanding of the scarce, highly qualified manpower which is also urgently needed in the civilian sector.

In the more immediate future, technological progress will—or at least should—influence the full spectrum of military operations envisaged in the Southern Region. As in the Northern Region, a key feature of operations in the Southern Region is the high degree of interdependence of land, air, and sea operations. To cite just a few examples, reinforcements for the land battle would mainly arrive by sea, entering the Mediterranean through the Straits of Gibralter; air operations over sea and land would involve land-based and carrier-based aircraft; and amphibious operations would be crucial in certain scenarios.

The naval balance in the Mediterranean unquestionably favors NATO. The Soviet Union has only a limited presence due to severe logistic constraints arising from the absence of basing facilities. Access to the region from the Black Sea is via the Bosphorous and Dardanelles, and from the Atlantic is via the Straits of Gibralter, so reinforcement would be problematic to say the least. Land-based aircraft such as the Backfire and the Blackjack armed with long-range anti-ship missiles could be brought to bear in the area but would have to pass through NATO airspace in transit. NATO, on the other hand, enjoys the presence of the formidable American Sixth Fleet as well as the naval forces of the Mediterranean Allies. The survival of the Soviet Mediterranean Squadron (SOVMEDRON) is therefore measured in days after the outbreak of hostilities.

Advances in naval technology, however, will certainly complicate maritime operations as both Soviet and NATO capabilities improve.

At present, SOVMEDRON generally stands at 40 to 50 vessels typically including one or two cruisers, three to eight destroyers and frigates, and six to eight submarines. The balance consists of auxiliaries and supply vessels which outnumber warships by about two to one. In a crisis, the fleet could be augmented by assets from the Black Sea and the Atlantic but even without reinforcement, SOVMEDRON could cause major difficulties for NATO naval assets.

Soviet naval forces are becoming increasingly potent as new systems are continually fielded. Soviet anti-ship missiles are extremely capable and it is worth noting that many are nuclear capable. Some Soviet submarines carry up to 24 anti-ship missiles and could render a pre-emptive attack on NATO surface assets extremely damaging. Even though Soviet surface vessels could subsequently be neutralized by surviving NATO naval forces and land-based aircraft, that would be poor compensation for the loss of NATO aircraft carriers and amphibious vessels needed to support the land battle.

Even if the Soviet Union could not achieve a large-scale, pre-emptive attack using its anti-ship systems, judicious employment of its submarine fleet—particularly its most capable elements—could still severely disrupt NATO's operations. Soviet submarines in the Mediterranean or in the approaches to the Straits of Gibraltar could pose a direct threat to warships and sea lines of communication through attacks with torpedoes and anti-ship missiles. They could also pose a troublesome indirect threat by covertly laying mines. At the very least, they could tie up NATO's naval assets in ASW and minesweeping operations thus delaying the use of these assets in support of the land battle. In particular, Soviet sub-surface operations in the western Mediterranean and the approaches to the Straits of Gibraltar could draw NATO's naval assets away from the eastern Mediterranean where they would be better placed to support the land battle.

Of course, Soviet naval forces do not have a monopoly on new technology. It is by no means clear, however, that Soviet advances in submarine technology are wholly off-set by Western progress in ASW technology. Consequently, in the future, Soviet submarines in the Southern Region can be expected to cause greater problems for NATO forces than in the past, even though they would eventually succumb to NATO's local naval superiority.

NATO's amphibious operations in the Southern Region will certainly benefit considerably from a variety of new technologies embodied in systems now in the early stages of production. JTIDS is one example already mentioned earlier but also noteworthy are new transport systems for amphibious forces. American marine forces are being equipped with an amphibious assault hovercraft, the Landing Craft Air Cushion (LCAC) which has several crucial advantages over the traditional landing craft it supersedes. The LCAC is much faster, for instance, so that amphibious forces can be landed more rapidly and the parent vessel can remain further from the landing area. These factors render the location of a landing difficult to predict while also permitting a much more rapid build up of forces ashore. In addition, the LCAC has a substantial over land capability which allows a force to be deposited directly some way inland.

Amphibious forces will receive a further boost in capability with the advent of the V-22 "Osprey". This tilt-rotor vehicle has the take-off and landing flexibility of a helicopter and the forward-flight characteristics of a fixed-wing aircraft. It is thus an ideal vehicle for amphibious operations with much better speed and range than traditional helicopters. It is also more robust, more fuel efficient, and easier and cheaper to maintain.

In the Southern Region, a conflict between NATO and the Warsaw Pact could include any or all of three possible land and air campaigns. One possibility is a Warsaw Pact campaign into Eastern Turkey either as part of a larger conflict between NATO and the Warsaw Pact or in an attempt to secure the area to protect Warsaw Pact forces moving into the Middle East. This theatre is some 2,000 kilometers from Thrace and lines of communication are either by sea, over difficult terrain by road, or via an inadequate rail system. Reinforcement would therefore be very difficult for NATO so, with the exception of some American tactical air support, Turkish forces would operate in isolation. In their favor would be the harsh climate and mountainous terrain which would compel an invading force to take a limited number of routes and mountain passes.

A second possibility is a Warsaw Pact campaign in Greek and Turkish Thrace in order to gain control over the Bosphorous and the Dardanelles. In this area, the lack of a coordinated military command compounded by inadequate air defense radar coverage would be disadvantageous to NATO.

A third possibility is a Warsaw Pact campaign against North East Italy. In this case, Warsaw Pact forces would first have to penetrate Austria or Yugoslavia (or both) so attacking forces would be degraded and lines of communication would be long and vulnerable. In addition, the Alps present a severe obstacle and would channel Warsaw pact forces into narrow areas such as the Gorizia Gap.

Altogether in the Southern Region, the border to be defended is 1,800 kilometers in length. About 1,000 kilometers of this is the Turkish Black Sea coast which could face Soviet amphibious assaults to support operations in Thrace or Eastern Turkey or to draw NATO forces away from these areas.

The numerical land and air balances of NATO and Warsaw Pact forces in the Southern Region unambiguously favor the Warsaw Pact. In terms of quality too the Warsaw Pact enjoys an overall edge but the heterogeneous nature of NATO and Warsaw Pact forces does lead to some local variations. A critical deficiency for NATO is its anti-armor capability. Although this is relieved to some extent by the Leopard tanks operated by Greece, Italy and Turkey, much hardware is obsolescent and stocks of munitions are in some cases lamentably low.

The fundamental problem is, of course, a severe shortage of resources. The material provided by the more prosperous Allies compensates in part but donations of out-dated equipment do not represent the optimal solution. A more creative approach would be the introduction of some of the new technologies which seem particularly suited to operations in the Southern Region.

The use of RPVs and remote sensors in the mountain passes of Eastern Turkey and North East Italy would be useful complements to many of the new anti-armor munitions now in prospect. The MLRS also would be a useful addition, a fact recognized already by Italy which, unlike Greece and Turkey, can afford to add this system to its inventory. Fiber optic guided missiles, once operational, will be high-technology, low cost systems which could hold considerable appeal for NATO's less advantaged member nations, as could air-launched systems such as the hyper-velocity missile. The relatively unglamorous but nevertheless highly capable new mines now becoming available also seem to hold promise for some Southern Region operations, especially those where attacking forces would be compelled to advance along some relatively confined routes. Judicious use of a mixture of sophisticated mines in these areas could blunt and slow down an advance considerably.

Some possibilities would be more attractive than others and, naturally, the form of acquisition—direct purchase, donation or co-production—could be tailored as appropriate. The essential point, however, is that the more advanced NATO nations are in the process of developing a broad array of relatively inexpensive, effective anti-armor systems which could well meet some of the most pressing demands of the less prosperous nations in the Southern Region. It is also worth underlining that the technological sophistication of such systems does not preclude the possibility of manufacture in the Southern Region. Turkey, for instance, is already gearing up to manufacture the F-16, Italy has in a relatively short time become an extremely advanced producer of defense equipment and Spain seems set to follow this example. It therefore seems appropriate to consider transfers of new, high technology systems to the Southern Region and to assess the trade offs between the acquisition of large numbers of unguided munitions in favor—partly at least—of some "high leverage" smart systems.

Conclusion

The incorporation of new technology into NATO's force structure clearly will play a critical role in NATO's efforts to prevent the conventional balance from moving even further in the Warsaw Pact's favor. The availability of new technology is fortunately not a problem; NATO has a veritable treasure trove of technologies which promise profound improvements in capability. It is by no means assured, however, that NATO will exploit those technologies as effectively as possible. The track record is disappointing in that although NATO has sustained credible, technologically advanced conventional forces, it has paid over the odds to acquire them. In future, in the face of inflexible defense budgets, rising acquisition and operations costs, and growing Warsaw Pact capability, the luxury of inefficient procurement should be out of the question. To avoid it, NATO must pay greater attention to first defining and then meeting Alliance force requirements instead of essentially merging national ones. Furthermore, those requirements should be formulated on the basis of a mission-oriented rather than system-for-system philosophy. And in meeting requirements, procurement authorities must change their yardsticks for assessing performance; reliability and ease of maintenance must feature as prominently as do traditional parameters such as speed, range, and accuracy.

The NATO nations have at least made the first crucial steps towards redressing their procurement problems. They have recognized that trends in procurement are alarming and could culminate in disaster if left unchecked. They have also identified the sources of their procurement problems and are aware of the general nature of the necessary solutions. They have even taken measures to increase defense cooperation and competition and to define overall Alliance force requirements. Nevertheless, considerable room for improvement remains and expressions of increased willingness, greater readiness, and greater discipline will not suffice: greater movement towards cooperation, competition and coordination is essential.

NOTES AND REFERENCES

1. Phillip A. Karber *High Technology on the Battlefield. New Technology and NATO's Operational Doctrine.* Paper presented at the conference on "High Technology Security and Economic Growth" organized by the United States Mission to NATO, the EEC, Belgium, and the OECD held in Brussels in February 1986.
2. *The Military Balance 1985-86,* International Institute for Strategic Studies, London 1985. See pages 17 and 170.
3. *The NATO Report,* 28 July 1986 p.7.
4. Caspar Weinberger *The Technology Transfer Control Program: A Report to the 98th Congress.* United States Department of Defense, February 1984, pp.4-5.
5. For a detailed analysis see Stan Woods *Weapons Acquisition in the Soviet Union,* Aberdeen Studies in Defense Economics (ASIDES), No. 24, Summer 1982.
6. As note 2.
7. Speech by Lord Carrington, Secretary-General of NATO, on "NATO and European Security" at the Centro Studi per Conciliazione Internazionale, Rome, 24 October 1986.
8. François Heisbourg "Alliance and Sovereignty—Aims and Limitations in Common Endeavours" *NATO's Sixteen Nations,* November 1986, pp.18-25.
9. Thomas A. Callaghan Jr. "The Structural Disarmament of NATO", *NATO Review,* June 1984, pp.21-24.
10. Frank Cooper "Affordable Defense: In Search of a Strategy" *RUSI Journal,* December 1985, pp.1-6.
11. Kelvin Coriolis "The True Cost of Aircraft in Service" *Jane's Defense Weekly,* 27 September 1986, pp.680-683.
12. Ibid.
13. David Greenwood "Strike Shallow" *Defense Minister and Chief of Staff,* March 1984, pp.9-13.
14. Jim Meacham "Technology by Other Means" in *Marching Forward,* a survey of defense technology appearing in *The Economist,* 21 May 1983.
15. Graham Warwick "ATF: Balance Tips the Scales", *Flight International,* 6 December 1986, pp.28-29.
16. Coriolis op. cit.
17. Ibid.

18. Wayne Biddle "How Much Bang for the Buck?" *Discover,* September 1986, pp.50-63.
19. Dr. Steven L. Canby "The Operational Limits of Emerging Technology" *International Defense Review,* June 1986, pp.875-880.
20. Frank Cooper *Preconditions for the Emergence of a European Common Market in Armaments,* Center for European Policy Studies, Brussels 1985, p.13.
21. As note 9.
22. Thomas A. Callaghan Jr. *The Future of Transatlantic Armaments Cooperation within NATO.* Paper presented at the 6th Armed Forces Communications and Electronics Association symposium, Brussels 23-25 October 1985.
23. Brian Field "Economic and Defense Resources: The Prospect" *NATO Review,* October 1985, pp.24-29.
24. Dr. W. Dekker *The Technology Gap: Western Countries Growing Apart?* Paper presented at the Atlantic Institute for International Affairs, Paris 5 December 1985.
25. "Reagan Curbs Hit US Electronics Sales Overseas" *Financial Times,* 16 October 1986.
26. Michael Feazel "Eurofighter Bidding Rules May Bar U.S. Companies", *Aviation Week and Space Technology,* 19 January 1987, pp.18-19.
27. Mark K. Moore "Electronics: The Linchpin of Modern Warfare" *Armed Forces Journal International,* September 1986, pp.76-86.
28. Ibid.
29. Bruce D. Nordwall "Electronics Will Gain Larger Share of Budget" *Aviation Week and Space Technology,* pp.227-229.
30. Mark K. Moore op. cit.
31. Ibid.
32. Glenn W. Goodman Jr. "Air Force and Navy EW Programs Jammed, but Countermeasures Are in Sight" *Armed Forces Journal International,* February 1987, pp.66-73 and "NATO Emphasizes Joint Efforts to Meet Soviet EW Threat", *Aviation Week and Space Technology,* 9 February 1987, pp.83-89.
33. Michael Ganley "Congressional EW Concern" *Armed Forces Journal International,* February 1987, p.52.
34. Brigadier A.L. Meier O.B.E. "BICES. A Central Region Perspective", *International Defense Review,* October 1986, pp.1445-1449.
35. John M. Porter "Electronic Warfare. Can Everyone Tune in on the Same Frequency?" *Armed Forces Journal International,* February 1986, pp.40-52.
36. Philip J. Klass "Radar Countermeasures" *High Technology,* August 1986, pp.56-63.

37. John M. Porter op.cit.
38. Stephen Broadbent "EFA and Beyond. A Giant Leap for Avionics Software", *Jane's Defense Weekly,* 28 March 1987, pp.560-562.
39. Jean-Paul Christy "Optical Fibers for Defenses" *NATO's Sixteen Nations,* December 1986-January 1987, pp.78-81.
40. "VHSIC Program Provides DoD State-of-the-Art Microchips" *Armed Forces Journal International,* January 1987, p.62.
41. Glenn W. Goodman Jr. "VHSIC a Potential Boom to Air Force R&M and Advanced Tactical Fighter" *Armed Forces Journal International,* January 1987, pp.58-64.
42. Ibid.
43. Edward H. Kolcum "Joint STARS Intensifies Reliance on Software in Development Phase" *Aviation Week and Space Technology,* 23 February 1987, pp.81-84.
44. *New Technology for NATO. Implementing Follow-on Forces Attack* Office of Technology Assessment, United States Congress, June 1987, pp.146-154.
45. Ibid.
46. Jeffrey M. Lenorovitz and Keith F. Mordoff "European Seek Interoperable Battle Reconnaissance Systems" *Aviation Week and Space Technology,* 7 September 1987, pp.77-85.
47. Peter Hellman "The Little Airplane" *Discover,* February 1987, pp.78-87.
48. George Leopold "US to Seek Four-Member Family of RPVs" *Defense News,* 13 June 1988, p.1,8.
49. As note 44. pp.160-164.
50. John D. Morrous "Congressional Pressure Prompts Order to Revive Anti-Radar Drone" *Aviation Week and Space Technology,* 3 August 1987, pp.84-85.
51. Tim Wrison "RPV Development Accelerates" *Jane's Defense Weekly,* 27 September 1986, pp.687-690.
52. Ibid.
53. For more information on trends in tactical communications see Don Gordon "Frequency Hopping Radios—Rediscovery of Tactical Communication" *Military Technology,* May 1986, pp.54-67; Gowri S. Sundaram "Trends in Tactical Communications" *International Defense Review,* May 1987, pp.619-622; John Williamson "Revolution in Communication" *Jane's Defense Weekly,* 14 February 1987, pp.242-244.
54. Doug Richardson "Electronic Warfare" in *Advanced Technology Warfare* Salamander Books, London 1985, pp.12-43; Jim Meacham "Talk Swiftly, Strike Sure" *Marching Forward,* a survey of defense technology

published by the *Economist,* 21 May 1986.

55. Glenn W. Goodman Jr. "ASPJ, Mark XV IFF, and JTIDS Programs Make Significant Gains" *Armed Forces Journal International,* February 1987, pp.78.82.
56. Dan Gordon "The JTIDS/PLRS Hybrid—A NATO Standard?" *Military Technology,* May 1987, pp.97-106.
57. Ibid.
58. Christy Campbell *Airland Battle 2000,* Hamlyn, London 1986, pp.103-104.
59. Siegfried Seiffert "Technical Interoperability of the Command, Control and Information System on the Central Front" *Military Technology,* May 1987, pp.92-95.
60. Brigadier A.L. Meier O.B.E. op.cit.
61. David J. Marcus "NATO Intelligence System Faces First Major Review" *Defense News,* 15 February 1988, p.1,28; David J. Marcus "European Project Excludes US, Canada" *Defense News,* 14 March 1988, p.1,34.
62. Campbell op.cit. p.93.
63. John F. Rybicki "Advanced Conventional Munitions and the Airland Battle" *Military Technology* October 1986, pp.32-38; Daniel Charles "Bringing SDI Down to Earth" Science, 14 August 1987, pp.713-715; "One Shot, One Kill: A New Era of 'Smart' Weapons" *US News and World Report,* 16 March 1987, pp.28-35.
64. Campbell op.cit. pp.100-103.
65. Charles op.cit. and Bill Sweetman "MLRS to Launch New Weapons" *International Defense Review,* November 1987, pp.1503-1510.
66. John F. Fybicki. op.cit. and "We'd Have Done It Anyway, Honest" *The Economist,* 7 November 1987.
67. Mark Hewish "Trigat Moves Into Development" *International Defense Review,* October 1987, pp.1367-1371.
68. Steve Connor "Ground Wars" *New Scientist,* 23 July 1987, pp.44-49; Kenneth Richardson "Tactical Missile Technology for the 1990s" *Military Technology,* October 1986, pp.213-222.
69. "US, European Firms Seek to Develop Fiber Optic-Guided Antitank Missiles" *Aviation Week and Space Technology,* 20 June 1988, p.111.
70. Debra Polsky "Army Alters Acquisition of FOG-M" *Defense News,* 29 February 1988, p.1,18.
71. Michael J. Witt "Ajax/Apilas Weapon System Demonstrated to Prospective Customers" *Defense News,* 12 October 1987, p.34.
72. Steve Connor op.cit.

73. Kenneth S. Brower "Fuel-Air Explosives A Blow to Dismounted Infantry" *International Defense Review,* October 1987, pp.1405-1407; "CATFAE Anti-Mine System in Trials" *Jane's Defense Weekly,* 12 September 1987, p.519.
74. James J. Townsend "Sensor Technology and Armored Vehicles" *Military Technology,* May 1988, pp.74-78.
75. See for instance "Experts Clash Over US Ability to Stop Soviet Tanks" *International Herald Tribune,* 31 July 1987.
76. "US Says New Uranium-Based Tank Armor is Impenetrable" *International Herald Tribune,* 16 March 1988.
77. Paul A. Chadwell "Liquid Propellant for Guns" *Armed Forces,* June 1987, pp.255-258.
78. T.A. Heppenheimer "Electromagnetic Guns" *Popular Science,* August 1987, pp.54-58.
79. "Army Eyes Anti-Tank Robot" *Defense News,* 10 August 1987, p.3 and 24, and George Leopold "DoD Program Yields Progress for Three Autonomous Systems" *Defense News,* 12 October 1987, p.30.
80. Fred Reed "The Electric Jet" *Air and Space,* December 1986/January 1987, pp.42-48 and Mike Wilson "Fly-by-Wire Shapes the New Fighters" *Jane's Defense Weekly,* 25 August 1984, pp.285-293.
81. Jim Meacham "Back to the Wild Blue Yonder" in *Marching Forward,* a survey of defense technology appearing in *The Economist,* 21 May 1983.
82. Mike Wilson, op.cit.
83. Bruce A. Smith "AFTI/F-111 Flight Tests Demonstrate Potential of Mission Adaptive Wing" *Aviation Week and Space Technology,* 24 November 1986, pp.40-41 and Julian Moscon "Mission Adaptive Wing" *Flight International,* 10 August 1986, pp.22-25.
84. Lon Nordeen "The Applications of Composites in Combat Fighter" *NATO's Sixteen Nations,* June 1986,pp.66-68.
85. Bill Sweetman "Advanced Composites for Military Uses" *International Defense Review,* November 1986, pp.1631-1638.
86. "Aluminium-Lithium Structural Panels Tested on F-15" *Jane's Defense Weekly,* 6 December 1986, p.1339.
87. Bill Sweetman "The New Wave in Aircraft Power" *International Defense Review,* June 1986, pp.765-770; Graham Warwick "The Power Behind EFA" *Flight International,* 18 July 1987, pp.28-29; Kevin B. Wilshere "Fighters for the '90s" *Military Technology,* June 1987, pp.24-35.
88. Special Advertizing section on Aerospace *Aviation Week and Space Technology,* 10 August 1987, p.94.

89. Jeffrey M. Lenorovitz "Voice Command Unit Enters New Tests" *Aviation Week and Space Technology,* 30 January 1984, pp.65-66.
90. Graham Warwick "Flying with Flir" *Flight International,* 14 December 1985, pp.26-28; Mike Witt "Night Air Attack" *Military Technology,* April 1987, pp.23-33; Ferdinando Sguerri "What Technology for the Future Fighter?" *Defense Today,* August 1986, pp.298-300.
91. Stephen Broadbent "Stealing a Lead in Accurate Navigation" *Jane's Defense Weekly,* 13 September 1986, pp.569-570.
92. Kevin B. Wilshere "Flying into the 21st Century", *NATO's Sixteen Nations,* June 1987, pp.88-91.
93. Glenn W. Goodman Jr. "ICNIA Program Demonstrates Its Potential for ATF, ATA, LHX" *Armed Forces Journal International,* March 1988, p.64.
94. Stephen Broadbent "EAP: a Step up the Ladder of Advanced Technologies" *Jane's Defense Weekly,* 21 March 1987, pp.510-511.
95. George Leopold "US Promotes PAVE Pillar as Bedrock for Blueprint of Future Aircraft" *Defense News,* 24 April 1987, p.5; and Dan Boyle "Pave Pillar" *International Defense Review,* September 1986, pp.1249-1253.
96. T.A. Heppenheimer "Spreading the Stealth" *High Technology,* December 1986, pp.58-59.
97. T.A.Heppenheimer "Stealth" *Popular Science,* September 1986, pp.74-79, pp.115-116.
98. "New Materials Promise Low Radar Reflectance" *Aviation Week and Space Technology,* May 1986, p.22.
99. Graham Warwick "Flight Unseen" *Flight International,* 3 October 1987, pp.37-39.
100. Doug Richardson op.cit.
101. Ibid.
102. Bill Sweetman "Against the Odds. Air-to-Air Combat in the 1990s" *International Defense Review,* May 1987, pp.581-588; Graham Warwick "ATF: Balance Tips the Scales" *Flight International,* 6 December 1986, pp.28-30; Stuart F. Brown "21st Century Superfighters" *Popular Science,* October 1986, pp.80-86, 112-114.
103. Kevin B. Wilshere "Fighters for the '90s" *Military Technology,* June 1987, pp.24-35; Mark Urban "Where Eurofighter Fails to Fit to Defense Bill" *The Independent,* 17 April 1987; Benjamin F. Schemmer "France's Rafale Fighter May Not Be the 'Piece of Dog Meat' US Official Charged" *Armed Forces Journal International,* January 1987, pp.24-25.

104. Graham Warwick "For Agile Read Active" *Flight International,* 9 November 1985, pp.29-31.
105. Bill Sweetman "New Technology for the Military Helicopter" *International Defense Review,* February 1986, pp.183-188.
106. L.M. (Jack) Horner "Helicopters and High Technology" *Military Technology,* October 1987, pp.80-91.
107. "Anti-Tank Helicopters" *International Defense Review,* May 1987, pp.603-613.
108. Graham Warwick "Subhunters Extend Their Horizons" *Flight International,* 31 October 1987, pp.25-29.
109. As note 107.
110. As note 107. and Mike Gaines "LHX-A Giant Leap" *Flight International,* 1 June 1985, pp.167-171; Robert B. Ropelewski "Army Revises LHX Schedule, Advances Operational Capability to 1995" *Aviation Week and Space Technology,* 24 November 1986, pp.21-22.
111. Debra Polsky "LHX Program Gets Panel's OK to Start Demonstration Phase" *Defense News,* 13 June 1988, p.3,53.
112. Mike Gaines op.cit. and Tom Donnelly "Appearance of Tilt-Rotor Aircraft in LHX Competition Seems Likely" *Defense News,* 11 May 1987, p.4.
113. Tim Wrixon "Versatile V-22 Osprey" *Jane's Defense Weekly,* 2 August 1986, pp.170-172.
114. Julian Moxon "V-22 Osprey: Multi-Service Workhorse" *Flight International,* 30 August 1986, pp.154-157;
115. As note 108.
116. Julian Moxon "V-22 Osprey Changing the Way Man Flies" *Flight International,* 14 May 1988, pp.22-30.
117. Len Famiglietti and Paul Beaver "'Vital' V-22 Rolls Out to Warm USMC Reception" *Jane's Defense Weekly,* 4 June 1988, p.1095.
118. Peter Middleton "Osprey 'Will Survive'" *Flight International,* 4 June 1988, p.6.
119. As reference 117.
120. Brendan M. Greeley Jr. "Darpa Foresees X-wing Aircraft Full-Scale Development by 1992" *Aviation Week and Space Technology,* 20 October 1986, pp.80-85.
121. "X-Wing Scheduled to Fly in October" *Flight International,* 22 February 1986, p.18.
122. George Marsh "Crossed Wings Fly Faster" *New Scientist,* 11 December 1986, pp.42-45;
123. Stephen Ashley "X-Wing Aircraft" *Popular Science,* July 1987, pp.48-49, 90-91.

124. Bill Gunston *Warplanes of the Future,* Salamander Books, London 1985, pp.146-147.
125. Graham Clark "US, West German Firms Team Up to Develop Highly Manoeuvrable Fighter" *Defense News,* 31 August 1987, p.10.
126. David M. North "USAF Focuses Development on Emerging Technologies" *Aviation Week and Space Technology,* 24 February 1986, pp.18-19; Stephen L. Thomson "The Big Picture" *Air and Space* April/May 1987, pp.74-83; L.F.E. Coombs "The Cockpit of the Year 2000 and Beyond" *Military Technology,* July 1987, pp.28-38; Bill Sweetman "Against the Odds: Air-to-Air Combat in the 1990s" *International Defense Review,* May 1987, pp.581-588; Mike Gaines "Thought Control" *Flight International,* 30 October 1987, pp.40-41.
127. Kenneth Richardson "Tactical Missile Technology for the 1990s" *Military Technology,* October 1986,pp.213-226.
128. Ibid.
129. Bill Sweetman "Air Warfare in the 1990s" *International Defense Review,* August 1986, pp.1055-1063; Bill Sweetman "Against the Odds: Air-to-Air Combat in the 1990s" *International Defense Review,* May 1987, pp.581-588.
130. "Matra Missiles for the 90s" Interview with Emile Durand appearing in *International Defense Review,* June 1986, pp.865-866.
131. Bill Sweetman "Against the Odds: Air-to-Air Combat in the 1990s" *International Defense Review,* May 1987, pp.581-588.
132. Stuart F. Brown "The Missile That Waits" *Popular Science,* October 1987, p.69.
133. Mark Hewish, Bill Sweetman, and Gerard Turbé "Air-to-Surface Weapons: New Technologies for Precision Guidance and Stand-Off Delivery" *International Defense Review,* pp.599-609; David Fulghum "Air Force Tests Hypervelocity Missile With Ability to Kill Without Warheads" *Defense News,* 25 January 1988, p.45.
134. Glenn W. Goodman Jr. "NATO Accord Reached on 7-Nation Standoff Weapons" *Armed Forces Journal International,* September 1987, p.30; Dan Beyers "Allies Pursue Common Missile Program" *Defense News,* 30 March 1987, p.1,42; Ezio Bonsignore "LR-SOM + SR-SOM + LOCPOD = M-SOW (With Any Luck)" *Military Technology,* March 1987, pp.18-24.
135. David A. Brown "Short Brothers Pushes Startreak For US Army Air Defense" *Aviation Week and Space Technology,* 6 April 1987, pp.68-69.
136. "Expensive, But Necessary: the PATRIOT Surface-to-Air Missile System" *Military Technology,* October 1984, pp.33-50.

137. For more details of surveillance and reconnaissance satellites see David Hobbs *An Illustrated Guide to Space Warfare,* Salamander Books, London 1986, pp.50-59.
138. See Robert Banks (Rapporteur) "The Exploitation of Space" North Atlantic Assembly, February 1987, pp.64-65.
139. Philip J. Klass "Defense Dept. Will Seek Funds To Expand Navstar Constellation" *Aviation Week and Space Technology,* 5 October 1987, pp.30-32.
140. Julian Moxon "Aerospaceplane: Nasa's Flame Rekindled" *Flight International,* 10 October 1987, pp.31-32.
141. "The Intelligent Ship: A Future Concept" *Jane's Defense Weekly,* 5 September 1987, pp.467-470.
142. "MK 41 Vertical Launch System" *Armed Forces,* April 1987, p.184.
143. Lt. Col. David Miller "Naval Warfare" *Advanced Technology Warfare* Salamander Books, London 1985, pp.162-193.
144. G.R. Villar "Weapon Development 1984: Sea" *Royal United Services Institute & Brassey's Defense Yearbook 1985,* London 1985, pp.177-195.
145. "NATO, Industry Groups Study Use of RPVs for Small Ships" *Aviation Week and Space Technology,* 14 September 1987, pp.95-97; "Navy Begins Final Testing of Pioneer Short-Range RPV" *Aviation Week and Space Technology,* 14 September 1987, pp.93-95.
146. William Matthews "Possible DARPA Sponsorship May Keep Blimp Project Aloft" *Defense News,* 28 March 1988, p.42; "Large Airship Design Continues" *Flight International,* 21 May 1988, p.21.
147. Mick Hasser "Airships Face a Military Future" *New Scientist* 24 September 1987, pp.38-40; "US Navy Comiders AEW Airship Options" *Flight International,* 24 November 1986, p.12.
148. "Inverse Synthetic Aperture Technology Aids Radar Identification of Ships" *Aviation Week and Space Technology,* 7 September 1987, pp.88-89.
149. G.R. Villar op.cit.
150. Ronald O'Rourke "Defending Against the Cruise Missile" *Military Technology,* September 1986, pp.64-74.
151. Richard Layne "Invisible Subs" *Popular Science,* July 1985, pp.72-75.
152. Stephen Eisenstadt "Navy Envisions $5 Billion ASW Minisub Fleet" *Defense News,* 14 December 1987, p.1, 24; "Robot Submarines" *Wall Street Journal,* 11 March 1986.
153. "Sea Lance Enters Full-Scale Development" *Military Technology,* October 1986, pp.244-245.
154. "Homing Torpedo Defense" *Jane's Defense Weekly,* 26 September 1987, p.701.

155. Nick Cook "US/UK Torpedo Agreement Soon" *Jane's Defense Weekly,* 4 June 1988, p.1087.
156. T.A. Heppenheimer "Staying in Touch with Subs" *High Technology,* March 1987, pp.66-67; David Wanwyn Jones "Sending Signals to Submarines" *New Scientist,* 4 July 1985, pp.37-41; "ELF Planned for Scottish Wood" *New Scientist,* 31 October 1985, p.17.
157. T.A. Heppenheimer "Signaling Subs" *Popular Science,* April 1987, pp.44-48.
158. Lt. Col. David Miller op.cit.
159. Jonathan B. Tucker "Cold War in the Ocean Depths" *High Technology,* July 1985, pp.29-38.
160. Ibid.
161. Tim Carrington "The Silent War" *The Wall Street Journal,* 25 June 1987; Michael J. Witt "Navy's P-3 Update Designed to Enhance Detection of Quieter Subs" *Defense News,* 14 December 1987, p.19.
162. Jonathan B. Tucker op.cit.
163. Ibid.
164. "Experimental Underwater Array for Testing by RN" *Jane's Defense Weekly,* 15 August 1987, p.228; Michal J. Witt "Fiber Optics, Lasers Form Sensor Base" *Defense News,* 14 December 1987, p.14.
165. Jonathan B. Tucker op.cit.
166. Doug Richardson op.cit.
167. "Strategies of ASW" *Jane's Defense Weekly,* 5 September 1987, pp.453-465.
168. Lt. Steven R. Southard "Run Silent, Run Deadly" *Popular Mechanics,* April 1987, pp.86-89, 164-170; and Jonathan B. Tucker op.cit.
169. E.R. Hooton "The British Torpedo: On Track at Last?" *Military Technology,* July 1986, pp.86-94.
170. Lt. Col. David Miller op.cit.
171. Barbara Starr "P-3C Replacement Plan Seen as a Move Toward Closing ASW Gap" *Defense News,* 5 October 1987, p.12.
172. Alessandro Turrini "Mine Warfare: Minelaying and Neutralization" *Defense Today,* April 1986, pp.161-169.
173. Lt. Col. David Miller op.cit.
174. Alessandro Turrini op.cit.; Robert King "US Navy Funds New Antisubmarine Advanced Sea Mine in UK" *Armed Forces Journal International,* November 1986, p.30.
175. *Soviet Military Power 1987* US Department of Defense, p.88.
176. Scott C. Truver "Airborne Mine Countermeasures in the US Navy's Front Line" *International Defense Review,* October 1987, pp.1353-1355;

Ben Kocivar, V. Elaine Smay, and Richard Layne "Super Sweepers vs. Smart Mines" *Popular Science,* January 1985, pp.52-55, 86-87.
177. Lt. Col. David Miller op.cit.
178. "NATO Needs Common R&D Funding Says Minister" *Jane's Defense Weekly,* 17 October 1987, p.869; Speech by Lord Carrington, former Secretary General of NATO at the International Business Outlook Dinner, Woodstock, 26 September 1986; Speech by Robin Beard at the 6th Armed Forces Communications and Electronics Association symposium, Brussels, 23-25 October 1985.
179. *Towards a Stronger Europe* A Report by an Independent Study Team for the Independent European Program Group, Brussels 1986, Volume I.
180. Benjamin F. Schemmer *Improving Armaments Cooperation: There's a New Deck of Cards on the Table'* Paper presented at the conference on High Technology Security and Economic Growth organized by the United States Mission to NATO, the EEC, Belgium and the OECD, held in Brussels in February 1986.
181. Decision Document of the First Ministerial IEPG Meeting, 22-23 November 1984.
182. *Strategic Survey 1985-1986* International Institute for Strategic Studies, London 1986, pp.38-42.
183. Ibid.
184. Mack F. Mattingly "Conventional Armament Cooperation: Putting the Cart Before the Horse" *NATO Review,* October 1987, pp.26-30.
185. David Buchan "US Competition Takes on a Dizzying Aspect" *Financial Times,* 14 July 1987.
186. Kenneth Freeman "Defense Procurement Policy in Europe: Competition, Industrial Policy and Restructuring" *RUSI Journal,* December 1987, pp.23-32.
187. As note 185.
188. Bruce George (Rapporteur) *Draft General Report on Alliance Political Developments* North Atlantic Assembly, May 1985, p.7.
189. "France, UK Link on Arms Purchases" *Financial Times,* 11 March 1987.
190. As note 186.
191. Beard op.cit. p.32.
192. "France, UK Discussing Reciprocal Navy Purchases" *Financial Times,* 26 March 1987; Lesley Bedford "UK/France Talks on Off-the-Shelf Buying" *Jane's Defense Weekly,* 26 September 1987, p.703; Sergio Rossi "Countries Work Toward Defense Cooperation" *Defense News,* 16 May

1988, p.3; Michael J. Witt "British, French Develop Rapport on Defense Procurement" *Defense News,* 13 June 1988, p.44.

193. Kenneth Freeman op.cit.
194. As note 183.
195. For a detailed analysis of technology transfer issues see Ton Struick Van Bemmelen (Rapporteur) *Interim Report of the Sub-Committee on Advanced Technology and Technology Transfer,* North Atlantic Assembly, September 1987.
196. David Greenwood op.cit.
197. Andrew J. Pierre (ed) *The Conventional Defense of Europe: New Technologies and New Strategies,* Council on Foreign Relations, New York 1986, p.21.
198. David Greenwood *New Defense Policies and Their Affordability.* Paper presented at *The Economist* conference on *A Transatlantic Debate over Emerging Technologies and Defense Capabilities: A Watershed for the Alliance,* London, 9-10 February 1986.
199. Ibid.

ABOUT THE AUTHOR

David Hobbs is currently the Director of the North Atlantic Assembly's Scientific and Technical Committee and an Honorary Visiting Fellow in Defense Studies at Aberdeen University. He obtained his first degree in Physics with Science and Technology Policy from Manchester University, and then took a Master's degree in War Studies at King's College at the University of London.

Before taking up his position at the North Atlantic Assembly, he was a researcher at Aberdeen University's Center for Defense Studies, where he wrote monographs on intermediate-range nuclear forces and on the modernization of the United Kingdom's independent nuclear deterrent force.

He has published articles in journals such as *Jane's Defense Weekly, Naval Forces,* and *Marine Policy* and has contributed chapters to several books. He has presented numerous conference papers on security-related technology issues and is the author of *An Illustrated Guide to Space Warfare* published by Prentice Hall (US) and Salamander (UK).